Learning by Doing

Science and Technology in the Developing World

About the Book and Author

Science and technology capabilities are crucial to the economic growth of developing countries and to their ability to compete in the world economy. What factors enable some countries to successfully adapt technology to create indigenous capabilities and what factors cause others to fail? In this first global survey of science and technology capabilities in developing countries, the authors examine the experiences of Africa, the Caribbean, Latin America, the Middle East, China, India, and East Asia. Specialists in science and technology policies in these regions emphasize learning by doing: using available science and technology in its various applications--the shop floor, universities, and research institutes--to eventually develop indigenous capabilities. The authors consider why such capabilities have emerged in some societies but not in others and discuss their importance for domestic and international relations. Also considered are the implications of the "learning by doing" process for international relations, international trade, regional studies, science and technology policy, and management studies.

This unique survey will interest a large audience, from technology policymakers and regional specialists to businesspeople, managers, and officials. It will serve as a reference guide to the current state of science and technology policies in every region of the world and as a framework for analyzing and understanding how science and technology capabilities are being developed.

Aaron Segal is presently on leave from the University of Texas at El Paso teaching at the Air War College. Formerly an administrator of International Programs at the National Science Foundation, he has published extensively on science and technology policies in developing countries.

Learning by Doing

Science and Technology in the Developing World

Aaron Segal

with Brijen Gupta,
Wallace C. Koehler, Jr., Ward Morehouse,
Richard P. Suttmeier, and Wenlee Ting

NEW YORK AND LONDON

First published 1987 by Westview Press, Inc.

Cover graphic by Sally M. Segal © 1987

Published 2021 by Routledge
605 Third Avenue, New York, NY 10017
2 Park Square, Milton Park, Abingdon, Oxon OX14 4RN

Routledge is an imprint of the Taylor & Francis Group, an informa business

Library of Congress Catalog Card Number: 87-062197

ISBN13: 978-0-3670-0618-1 (hbk)
ISBN13: 978-0-3671-5605-3 (pbk)

DOI: 10.4324/9780429036040

Contents

Tables and Figures

TABLES

FIGURES

Contributors

Brijen Gupta is the Research and Development Director of the Council on International and Public Affairs and co-author of A Study of Indian Science and Technology Relations (1980). He has taught at the University of Rochester and elsewhere.

Wallace C. Koehler, Jr., is Head of Technology Assessment at the Center for Energy and Environment Research at the University of Puerto Rico. He has written extensively on energy and science and technology policy issues.

Ward Morehouse is President of the Council on International and Public Affairs, Chairman of the Intermediate Technology Group of North America, and a Research Associate in the Southern Asia Institute of Columbia University. His publications include several books on science and technology in India and studies of the social impact of technology on the Third World.

Aaron Segal is Professor of Political Science at the University of Texas at El Paso and a Visiting Professor at the Air War College of the U.S. Air Force. Formerly a manager of international programs at the U.S. National Science Foundation his research experience and publications include Africa, the Caribbean, and Latin America.

Richard P. Suttmeier is Henry P. Bristol Professor of International Affairs at Hamilton College in Clinton, New York. He is the author of two books and numerous articles on Chinese scientific and technical developments.

Wenlee Ting is an Associate Professor at the American Graduate School of International Management in Arizona. He is the author of Business and Technological Dynamics in Newly Industrializing Asia (Greenwood 1985), publisher of a newsletter, Asian Business Insider, and President of Transource Services.

Acknowledgments

This book is the product of many influences and individuals. While the authors alone are responsible for what appears on the printed page it is important to recognize the contributions of others. Deborah Lynes, Kate Frew, Jody Shulins, and Kellie Masterson at Westview Press were patient, encouraging, and stalwart editors. Vickie Gilbert provided stellar word processing services.

My own odyssey into the world of science and technology includes valuable stages at different times at the Science, Technology and Society Program at Cornell University, the Division of International Programs of the National Science Foundation, and the Air War College of the U.S. Air Force. A year spent in Mexico directing the Institute of Urban and Regional Development in Toluca provided a hands-on experience of applied research in a developing country.

Association with colleagues has been a primary learning source and I am especially intellectually privileged to have enjoyed the stimulation of Franklin Long and Dorothy Nelkin at Cornell, Henry Birnbaum and Carl Schwarz in Mexico, and Gordon Hiebert, Mildred Bosilevac, Bodo Bartocha, Eduardo Feller, Jean Johnson, and others at NSF. Margarete Luddemann has been a regular source of insights and information.

Finally, the emotional support that makes an effort of this duration possible has come from my wife, Sally, and my children, Janna and Marcus.

Aaron Segal

1

Learning by Doing

Aaron Segal

Many are called but few are chosen. Almost all of the 162 independent nation-states in the world have governments committed to the objective of developing indigenous science and technology capabilities. However research and development remains concentrated in North America, Western and Eastern Europe, Japan, and the Soviet Union (Tables 1.1 and 1.2). Only a handful of countries have been able to successfully proceed from the transfer of technology to the establishment and extension of indigenous science and technology generating capabilities.

This chapter examines what is involved in a given country or group of countries making the transformation from being importers of technology to having local science and technology strengths. It asks how the technology transfer process can be used to foster indigenous capabilities. It looks comparatively at the experiences of many developing countries, which are considered in detail in later chapters. The raw material consists of the failures and successes of many countries over lengthy periods of time.

Reprinted by permission of the publisher from The Political Economy of International Technology Transfer, John R. McIntyre and Daniel S. Papp, eds. (Quorum Books, a division of Greenwood Press, Inc., Westport, CT, 1986), pp. 95-116.

TABLE 1.1
Global Distribution of Research and Development Capacity, 1973

Region	Funds (billion dollars)	Share of World Total (percent)	Scientists, Engineers in R & D (thousand)	Share of World Total (percent)
Developing countries	2.77	2.9	288	12.6
Africa (with South Africa)	0.30	0.3	28	1.2
Asia (without Japan)	1.57	1.6	214	9.4
Latin America	0.90	0.9	46	2.0
Developed countries	93.65	97.1	1,990	87.4
Eastern Europe (with USSR)	29.51	30.6	730	32.0
Western Europe (with Israel and Turkey)	21.42	22.2	387	17.0
U.S.A. and Canada	33.72	35.0	548	24.1
Other (with Japan and Australia)	9.01	9.3	325	14.3
World total	96.42	100.0	2,279	100.0

Note: More recent global data is not available.
Source: Jan Annerstedt, Worldwatch Papers 31, Washington, DC, 1979

TABLE 1.2
Research and Development in the Western Hemisphere, 1980

Country	Researchers (full-time equivalent)	R & D Expenditures (US $ million)	R & D (% of GDP)	Technology Exports as % of Total Exports
USA	600,000	$ 65 billion	2.5	60
Canada	35,000	$1.2 billion	1.1	30
Brazil	12,000	$800 million	0.8	25
Mexico	8,000	$600 million	0.7	15
Argentina	7,000	$400 million	0.6	20
Venezuela	3,000	$250 million	0.4	5
Colombia	2,000	$ 90 million	0.3	15
Chile	2,000	$ 75 million	0.2	15
Cuba	1,500	$ 50 million	0.25	5

Note: The other 20 independent states of the Western Hemisphere each have 500 or fewer researchers, spend $10 million a year or less on R & D with these expenditures accounting for less than 0.1% of gross domestic product, while technology exports are 0-20% of total exports. The science and technology problems of the smaller states are thus of a fundamentally different nature. This table represents very crude estimates based on data from national science and technology plans, OECD, UNESCO, and National Science Foundation sources. There is little data on "shop-floor" industrial R & D in Latin America and it is probable that private sector R & D has been underestimated.
Source: Aaron Segal, Science, Technology and Western Hemisphere Governance, Aspen Institute Background Paper, 1982.

The methodology used is historical, empirical, and cross-cultural. It benefits from an extensive literature on both technology transfer and domestic science and technology (S & T) capabilities.[1] It utilizes available case studies of national experiences and seeks explicitly to extract from these diverse data some cross-cultural generalizations for which more than national applicability may be claimed. It is subject to the pitfall that national experiences are inherently sui generis, so that what worked in Japan may be irrelevant to India or vice-versa. However, it does not posit that there is one superior road to take in going from importing technology to achieving local S & T strength. It suggests, though, that there are several learnable dos and don'ts for policy-makers no matter what road they choose.

The transfer of technology is a constant of human history. Most societies have been at different times net exporters and net importers of technology. However, since the eighteenth century, Western Europe and then North America and later Japan have been predominantly exporters while Africa, Latin America, the Middle East, South Asia, and the Far East have been importers. This imbalance has been the direct result of the exporters being the first to develop domestic S & T capabilities and to sustain them. Any redressing of the imbalance will be a direct function of transfer leading to institutionalization in other parts of the world. Economic historian Walt Rostow has attempted to provide dates for industrialization for different countries (Figure 1.1). One need not agree with these exact dates in order to recognize the existence of first, second, late, and probably late-late comers to industrialization and domestic S & T. Similarly, the geographic diffusion within countries of transferred technologies leading to local institutions has been extremely uneven. India is only one of many countries currently in which high and rudimentary technologies and institutions coexist in close proximity as was the case in nineteenth-century industrializing Europe and North America.[2]

Definitions

There are many ways to determine the existence of a domestic S & T capability. This study emphasizes five empirical tests.

1. Absolute and proportionate expenditures on research and development (R & D), civilian and military. In contemporary terms an investment of $100 million a year and

FIGURE 1.1
Stages of Economic Growth: Twenty Countries

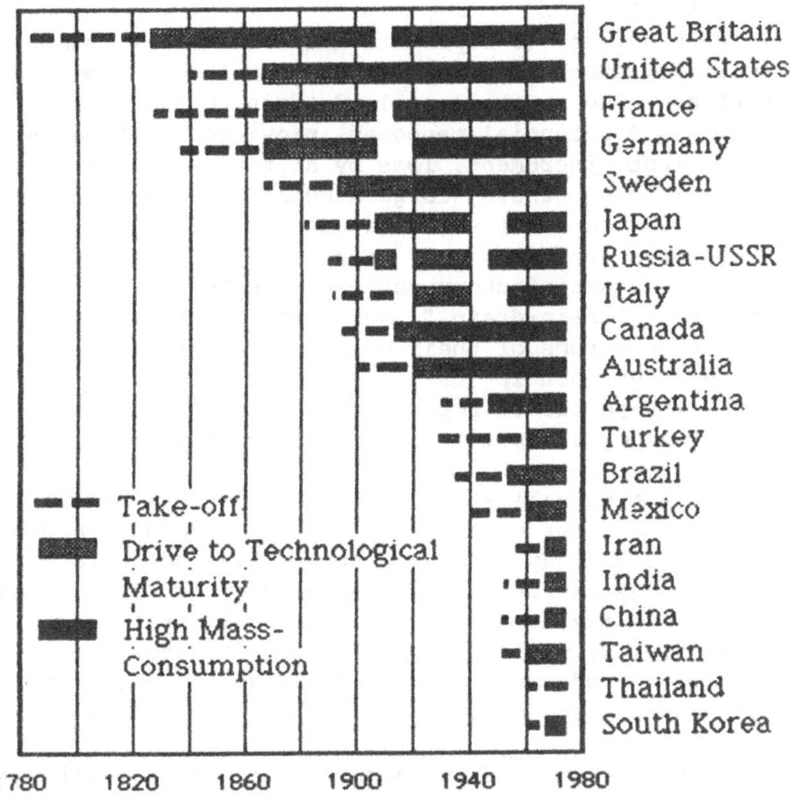

Source: W. W. Rostow, The World Economy: History and Prospect (Austin: University of Texas Press, and London: The Macmillan Press Limited, 1978), p. 51 and Part Five. Reprinted with permission of the publisher and the author.

.005 percent of gross domestic product is probably the minimum needed for a national capability. Smaller sums devoted to individual sectors may produce results while leaving a country overwhelmingly dependent on the technologies of others. However, no absolute or relative amount of spending can guarantee a capability, and the pursuit of such numerical targets as proposed by UNESCO and others is an illusion.

2. The ability to replicate and regularly increment the total national stock of scientists and engineers from local institutions. Due to the long lead times involved in producing R & D potential manpower, provision must be made for replacement, increment, loss by attrition, and other factors. Although sending large numbers of students abroad has been relied on for several decades by Japan, China and other countries, this can only be a means to an end and one that has many unsatisfactory outcomes, including the possibility of "brain-drain." While all countries will continue to send some of their best people for overseas study, a sine qua non is that local institutions can produce enough quality people to replace and expand the manpower pool. Many important policy implications stem from this condition.

3. Institutionalization is achieved when a net importer of technology becomes an exporter of science and technology even if the technology trade balance remains unfavorable. These technology exports should represent local design and engineering and may be distributed by local firms, multinationals, or various combinations. Assembly-plant offshore exports of imported components do not meet this criterion unless and until they are locally produced through backwards linkages.

4. Institutionalization occurs when national scientists and technicians make a contribution to the world stock of knowledge through copyrights, patents, licenses, presentation of papers at international meetings, publications in international journals, and other means. Countries may remain net importers of knowledge, as is the case for most of the smaller developed countries, but they need to possess an export capability.

5. The existence of institutionalization is indicated by applied research by nationals directed at _in situ_ problems which cannot be fully studied from abroad. Such applied research capabilities usually involve research institutes, interdisciplinary teams, and sustained long-term funding and management directed at agriculture, hydrology, forestry,

ecology, geology, entomology, animal husbandry, and other generically *in situ* problems. The lack of such capabilities is reflected in the import of research and researchers from abroad to tackle problems best studied at home.

Further Definitions

The five criteria mentioned above are the yardsticks for measuring the extent of domestic formal S & T capabilities - funds, manpower, trade and information flows, and applied research. Another set of criteria is needed to measure informal capabilities. It is likely that these informal strengths and weaknesses are more important to the achievement of a domestic capability than are the formal criteria. However, they are more difficult to detect and quantify.

These informal ways of acquiring domestic capabilities have been masterfully analyzed by economic historian Nathan Rosenberg.[3] They have been used by each of the nineteenth-century industrializers and have been documented as being used by the late and the late-late comers. Because they are often "shop-floor" practices they do not always show up in official figures of national R & D. They are the following:

1. Learning by doing. Normally at the manufacturing stage after R & D has been completed, this involves developing increasing skill in production and thus reducing labor costs per unit of output. Learning by doing does not require a highly trained and expensive scientific establishment but an entrepreneurial and innovative shop-floor atmosphere.

2. Learning by using. This may involve reverse engineering to take a capital good or other technology apart. Learning by using is often promoted by constraints on access to foreign exchange, shortages of spare parts, possibilities of economic savings through improved maintenance, simple embodied design modifications, or disembodied alterations in use.

3. Minor technological improvements. These often involve cost-saving improvement engineering.

4. Cost-reductions through improved maintenance and reliability.

5. Technological improvements that generate better science. This is particularly the case of instrumentation, optics, etc.

6. Growing interindustry relations as local contracts and subcontracts increase quality controls to meet new demands.

The work of Louis Wells, Jr., on Third World multinational firms and the detailed study of informal R & D in Argentina by Jorge Katz, as well as other research, confirm the importance of these informal criteria.[4] Their absence is the best evidence that indigenous capabilities do not exist, no matter what the formal figures may be for R & D spending. Ready availability of foreign exchange, use of turnkey projects, reliance on expatriate research personnel, inability to maintain machinery or to produce spare parts locally are among the many signs that the transfer of technology has not been institutionalized in spite of massive expenditures.

Transfer Capabilities

If so many societies are called to science and technology why is it that so few are chosen? Rosenberg holds that

> One of the most compelling facts of history is that there have been enormous differences in the capacity of different societies to generate technical innovations that are suitable to their economic needs. Moreover, there has also been extreme variability in the willingness and ease with which societies have adopted and utilized technological innovations developed elsewhere. And, in addition, individual societies have themselves changed markedly over the course of their own separate histories in the extent and intensity of their technological dynamism. Clearly, the reasons for these differences, which are not well understood, are tied in numerous complex and subtle ways to the functioning of the larger social systems, their institutions, values, and incentive structures.[5]

The ability to effectively transfer technology is essential to the achievement of a domestic capability, but it will not guarantee such a capability. The transfer of technology has for many societies served as a catalyst for the transformation of domestic institutions. Each industrializing society in its turn has begun at the lowest rung of technological complexity and moved up the ladder as its capabilities grew. In this way industrial technology has been globally diffused for nearly 150 years, with little control over its spread. Multinational corporations, joint ventures, local firms, public-sector enterprises, multinational and local banks, financing houses, and other institutions have been used to transfer technology.

What has enabled some societies to successfully transfer technology while fostering domestic capabilities while other countries have faltered? Rosenberg argues that "perhaps the most distinctive factor determining the success of technology transfer is the early emergence of an indigenous technological capacity."[6] He cites the early Japanese insistence on local control of technology and their adaptation to reduce capital-output ratios in order to take advantage of their small-scale, labor-intensive traditional industries. Korea, Taiwan, and other late-comers have also relied on capital-stretching adaptations of imported technology to better fit local factor ratios. Rosovsky finds in the Japanese experience of improvement engineering that "simple and small improvements suited to local conditions are frequently possible."[7] The analyses of third-world multinationals reiterate their cost-competitiveness through better adaptation and maintenance of proven technologies. These empirical studies are consistent with the concept of technological progress as being incremental, gradual, and through many small modifications rather than being based on major breakthroughs. The existence of broad, popular education has probably been more important to the industrializers than the presence of a few scientific geniuses.[8]

The transfer of technology is then for Rosenberg "an ongoing activity. The successful transplantation of a technology involves the domestic capacity to alter, modify, and adapt in a thousand different ways." The Japanese success is based on their having emphasized indigenous capabilities from the outset. The lesson of their success "seems to have been a government strategy for introducing foreign technologies in ways that emphasized their local linkages and the emergence of an indigenous technological capacity."[9] However, the emergence of science-driven technology and the predominance of multinational firms with centralized management and research may constrain local learning by doing and using and other informal skills. Can late-late comers lacking strong science capabilities find high-technology niches in which they can compete? Will the product-cycle continue to favor the adapters and improvement engineers who lack the R & D admissions price?

Can technology transfers be structured so as to foster indigenous capabilities? The question is obvious but the answers are not. Most technology transfers are commercial transactions based on short-run cost considerations. Where technology transfer is used to provide import-substitution, nationalism and cost combine to override other factors. The need to conserve foreign exchange, to acquire advanced weapons systems, to undertake major projects, and other factors all outrank domestic S & T capabilities. National communities of researchers are often politically suspect or weak and poorly placed to make their views felt, especially if they are involved in academic or basic research. Technology transfer simply is not often conceived of as a means of strengthening local capabilities. Much learning by doing and using is the unanticipated response to technology transfer rather than its planned outcome.

There is no clear relationship between many of the proposals for modifying technology transfers and the development of domestic capabilities. This is because these proposals are intended for other purposes; primarily to improve the negotiating position of developing country governments or firms vis-a-vis multinational corporations. Thus proposals for joint ventures, limited partnerships, the unbundling of technological packages, national and international regulation of royalties, licenses, copyrights, patents, and fees, and other regulatory measures are essentially redistributive in nature.[10] There is little evidence that they contribute to local R & D, whether performed by multinationals or national firms. Instead they tend to proliferate national technology bureaucracies which perform registration and regulation functions.

National legislation to screen imported technologies has been developed in India, Mexico, Brazil and elsewhere. Although its implementation in each society has been strikingly different the overall pattern is one in which import-substitution has been the goal rather than technological adaptation or innovation. Brazil has used limits on technology transfer to reserve for local firms and joint ventures the micro-computer market which has generated some R & D, especially of software.[11] India's extensive pursuit of technological self-reliance has spurred some local research in chemical fertilizers and other fields but also much bureaucratic inefficiency.[12] There is little evidence that technology transfer regulations have been used nationally to screen for more "appropriate technologies," partly because technology transfer is done by bureaucrats lacking research capabilities. How are two economists in the Dominican Republic to decide which 5,000 requests to import drugs are appropriate? International organizations have done little to provide the kind of back-up information services that would be needed if national technology transfer mechanisms were to actually choose between technologies.

Foreign exchange rates may play a more important role in promoting domestic capabilities than formal technology transfer mechanisms. Overvalued currencies and easy access to foreign exchange encourage reliance on foreign "trouble shooters," spare parts, and consultants and discourage learning by doing and using, improvement engineering, and local maintenance. The oil-exporting countries have been conspicuous for the weakness of their domestic capabilities. It is too easy to call Zurich or New York and have someone come with a part. Conversely reliance on used and proven technologies, by Chinese entrepreneurs in Indonesia and elsewhere, has stimulated local adaptation.

It is possible to suggest what forms of technology transfer are best and worst for domestic capabilities. Multinational corporations (MNCs) which transfer proven technologies for protected overseas markets discourage local learning or R & D. Local firms including public sector enterprises which license proven technologies for highly protected national markets also have little incentive to adapt. Bundled technological packages which leave decision-making to the provider discourage local capabilities.

Arms-length transfers between multinationals from rich countries and local firms have served in India, Taiwan, South Korea and elsewhere to encourage innovation. Joint ventures for export from developing countries which emphasize quality control have had similar effects. The rule seems to be that rich-country multinationals should either have a direct economic stake in local learning or else be sufficiently distant so that local firms are free to tinker.

MNCs have shown little interest in establishing developing country R & D facilities, although skilled researchers are sometimes available at low labor costs.[13] The size of their markets and approaches to management usually dictate centralized research in a few locations. However few developing country governments or firms have sought to tie technology transfer to local R & D, whether through licensing, fiscal incentives, or other approaches. The obvious sector in which to begin is agriculture, especially agricultural implements and fertilizers where downstream applied research costs may be low. At present petroleum, gas, and mineral exploration are the principal R & D expenditures by multinationals in developing countries. This extractive research is often performed with minimal local participation.

Transferring Capabilities

Transfers of technology which are specifically intended to build institutional capabilities in science and technology belong in another category. These have a long and distinguished history including European missions to Egypt and to the Ottoman Empire in the nineteenth century, the foreign medical schools established in China in the early twentieth century, the American Protestant-founded American University in Beirut, and Roberts College in Istanbul, and many others. During the last decade the World Bank and other donors have concentrated on institution-building in national agricultural research capabilities to network with the eleven international agricultural research centers. Although the developing country experience with science and technology institution-building through external transfers is massive it has received relatively little attention. Often it is lumped together with other external aid transfers and its significance is lost.

Again there has been over-emphasis on transfer and inadequate attention to domestic institutional innovations. External aid has funded several dozen African versions of French and British universities with scarce institutional innovation. India has done much better at transferring with external aid institutions modeled after MIT or LSE (London School of Economics) and adapting them.[14] Latin America has found institutional innovation in nonacademic research institutes supported by local foundations and national governments. Latin American public universities steeped in internal conflicts have resisted innovation. Family-owned firms in Taiwan, South Korea, Hong Kong and elsewhere in Asia have proven adept institutional innovators while their counterparts in Latin America and the Middle East have floundered.[15] No doubt shop-floor learning does better in certain management environments than in others.

Institutional capabilities are much more difficult to transfer than specific technologies. Scores of developing countries have had telephone equipment transferred. Few have been able to maintain the equipment; not to mention learn from using it. The World Bank in a paper on agricultural research contends that

> The country without the capacity to carry out research on its own benefits very little from research done elsewhere. A developing country's ability to screen, borrow, and adapt scientific knowledge and technology requires essentially the same research capabilities as those needed to generate new technology. Yet, few national systems have so far developed the administrative and technical capabilities to absorb and adapt, in an effective way, knowledge and technology that is becoming available to them from work at the international centers and research institutions in developed countries.[16]

The eleven international centers with a total annual budget over $100 million train national scientists but do not otherwise assist institution-building. Unlike industrial technology transfers where regulation can be provided by bureaucrats, agricultural effective transfers require national scientists and technicians who are also needed for research.

Technology transfers, whether through education and training abroad or implantation of models of developed country research institutions overseas, are problematical means of promoting local capabilities. Individuals and ideas travel better than institutions. Individuals who have studied abroad may be effective "change agents" but not necessarily institution-builders. Two generations of foreign educated Chinese scientists were able to survive the Cultural Revolution and keep science and technology alive but unable to train their successors. Instead China has had to dispatch another generation of scientists abroad to provide leadership for its own struggling institutions. Egypt, Iran, and Turkey are examples in which a reverse flow of scientists and engineers occurs while national institutions falter. External efforts at institutional-building for science and technology have a long list of failures. The international agricultural research centers and other similar efforts have been carefully insulated from the national environments and exist as scientific oases poorly networked to local staff.

The World Bank has underlined the failure of agricultural research transfers to take.

> National research programs are typically the weakest links in the global research effort. Common deficiencies include excessive fragmentation of research activities among government agencies, low priority accorded to research by governments, and inadequate institutional structures for research and extension. External assistance to strengthen national systems must take into account the size of the country's agricultural sector and the current state of development of its research system. Perhaps ten percent of developing countries already have adequate research skills, good national research programs, and effective linkages with international research institutions.... Another ten percent of the developing countries have adequate research expertise, but it frequently is poorly organized and managed. For this group of countries, external assistance in research organization and management may be required. Nearly half of the developing countries are large enough to justify and support a balanced national research system but lack essential research infrastructure. The needs of these countries are to develop an effective organization for research, to acquire proper research facilities, and to strengthen the scientific manpower base to conduct research. The remaining countries have very limited research resources and no single crop of sufficient importance to warrant a complete research system. For these countries, the major need is to develop a limited capability for research, largely of an adaptive nature, for a small number of economically important crops.[17]

The mixed record of the World Bank and other donors in transferring agricultural research capabilities compares favorably with the attempts of international organizations to transfer science and technology policy planning capabilities. While China, India, Brazil, Argentina, Mexico and a few other developing countries initiated their science and technology planning efforts in the 1960s and earlier, UNESCO and the United Nations Conference for Science, Technology, and Development (UNCSTD) gathered momentum in the 1970s. The apogee of this effort was the 1979 UNCSTD Conference in Vienna at which all attending developing governments were to submit a statement of national science and technology policies. A small army of consultants, tons of manuals, and numerous commissioned studies by international organizations stressed the importance of science and technology to development, and the need for national policies and planning capabilities.

The fall-out from this international consciousness arousing exercise was limited. The inventories of national capabilities that resulted in response to international pressure was for many countries a useful exercise even if much of the data obtained is questionable. However the Conference quickly turned to regulating technology transfer where views were polarized and skipped over national capabilities except to seek more external aid. Only those governments which had committed themselves to planning S & T prior to the 1979 Conference continued to do so afterwards. The highly centralized kind of planning with an emphasis on technology transfer regulation which was advocated by UNESCO and UNCSTD is impractical for most countries, and had few takers.

Science and technology policy is an excellent example of the problems of institutional technology transfer. While the S & T experience of the US, the USSR, the United Kingdom, Japan, France, Sweden and other developed countries is of considerable interest and some relevance it is not applicable or transferable. Where it is taken as a model as in the Cuban effort to replicate Soviet centralized science and technology, the result has been poor. Science and technology policy is as much _in situ_ as is research on food crops. There is more transferability between the experience of Mexico with a decade of S & T planning and Ecuador which is just getting started but there is still no model. The attempt of several international organizations to export a model for S & T policy and planning failed and not much learning by doing has followed. The comparative study of science and technology policy in ten developing nations funded by the Canadian International Development and Research Center was a much more careful and useful empirical effort.[18] It too found that economic variables such as exchange rates and fiscal policies were more important to promoting domestic S & T capabilities than national planning.

Rating Domestic Capabilities

A very crude rating of the current domestic S & T capabilities of individual developing countries can be derived by using these formal and informal criteria. The rating also includes an assessment of a country's ability to effectively absorb and adapt transferred technology and technology related institutions. The ratings offered in the chapter (Table 1.3) are arbitrary, partial, sketchy, based on extremely limited data and sources, and deliberately impressionistic. No attempt has been made to weight or rank order the criteria used for rating. What is intended is a snapshot at a particular moment of time of where individual countries are in a race which extends historically backwards and which has no specified finish.

Moreover these are ratings of domestic capabilities which have been achieved and are likely to be permanent; the very essence of institutionalization. The ratings do not reflect the objectives to which a country may put its hard-won domestic capabilities. Thus Brazil uses its indigenous design and engineering to export armored cars and civilian jets while research on domestic food crops lags. Similarly India has achieved impressive space, nuclear energy, and sophisticated weapons technologies while the majority of its population has remained at a subsistence level. The objectives of maximizing employment, improving income distribution, or devising appropriate technologies cannot be achieved without the attainment of a strong indigenous capability. However that capability need not and often is not used to pursue these objectives. The individual country ratings offered in Table 1.3 do not correlate closely with other equity goals.

India is the only developing country which rates as fully institutionalized.[19] It is a substantial exporter of locally designed industrial products through its own and other multinationals, it has excellent universities and research centers and is capable of expanding and replacing the third largest research community in the world, its contribution to the world stock of scientific information is impressive, and its capacity for learning by doing, institutional innovation, and other forms of local adaptation is proven. India's weakness lies in its inadequate support

TABLE 1.3
Developing Country Science and Technology Capabilities, 1984

Fully Institutionalized	Semi-Institutionalized	Partly Institutionalized	Partly Institutionalized, Petroleum Dependent	Longshots	Apparent Failures	Others: Nonstarters
India	Brazil	Argentina	Algeria	Barbados	Chile	130 Developing Countries
	China	Malaysia	Iran	Colombia	Cuba	
	Hong Kong	Mexico	Iraq	Costa Rica	Egypt	
	Singapore	Pakistan	Kuwait	Jamaica	Indonesia	
	South Korea	South Africa	Saudi Arabia	Sri Lanka	Nigeria	
	Taiwan		Trinidad and Tobago	Thailand	Philippines	
			Venezuela	Turkey	Vietnam	

for and interest in low-cost, labor-intensive technologies. Its domestic capabilities have impacted mostly on the urban middle and upper-income centers and resource-favored agricultural areas such as the Punjab.

The semi-institutionalized countries are close to achieving sustainable permanent capabilities but still face major obstacles. Due to uncertainty over its political future when the British lease ends in 1997 Hong Kong has not invested in a science-based R & D capability and major research centers. Taiwan and Singapore have more secure political futures but also must make the transition from technology adaptation to acquiring major science-driven technology capabilities. China is struggling to replace a senior generation of foreign-trained scientists and has lost a generation due to the Cultural Revolution barriers to research. China also is groping to find management and organizational structures that will promote learning by doing and using and manufactured exports. China's strength lies in its experience with appropriate technologies, its widespread literacy, and its ability to diffuse popular science.[20] Brazil has made great strides in exporting locally designed products, has improved its academic and nonprofit research but still has problems producing all the quality manpower it needs. Its other weaknesses include lack of support for appropriate technologies, highly protected import-substitution industries with poor learning capabilities, and limited user ties to researchers. Brazil's greatest strength perhaps lies in its _in situ_ research capabilities for Amazon Basin research, hydrology and hydropower, and geology, alternative energy sources, and other topics.[21]

The partly institutionalized countries have all achieved significant advances in capabilities but remain subject to possible reverses or stagnation. The prognosis for each country depends on many variables and further progress is uncertain.

Argentina with its democratically elected regime in 1983 is seeking to redress twenty years of damage to what once were the finest capabilities in Latin America. What remains after a massive brain-drain is a pool of researchers which cannot replace itself, battered research centers and universities, some good science, a limited ability to adapt imported technology and to export local designs, and a tremendous job of rebuilding.[22]

Malaysia has done well at rubber and other agricultural research and has begun the process of technology adaptation and some industrial export. Persons of Chinese and Indian origin predominate among its entrepreneurs and researchers and the Malay-controlled government is ambivalent about the role of the minorities and the private sector in S & T.

Mexico has shown some ability to export local technology and to do solid applied research. However low academic standards require that researchers often be partly trained abroad and there are severe shortages of qualified manpower. Public-private sector, user-researcher links are poor and much academic and public-sector research is barely diffused. Domestic capabilities are strongest in the state-monopoly petroleum and petrochemical sector which is highly bureaucratized and limited in its ability to learn by doing or using. Multinational corporations do little R & D in Mexico and government technology transfer regulations do not offer incentives.[23]

Pakistan has achieved the greatest S & T capability of any of the forty Islamic states. It is strong in several scientific areas, has limited technology export experience, and is good at the learning skills. Its weakness lies in lack of inter-industry and public-private sector ties, neglect of appropriate technology and to a lesser extent agriculture, and a brain-drain, to the Middle East and the West.[24] Given political stability Pakistan has a good chance of moving up to the semi-institutionalized status provided that it can keep good people at home.

South Africa with 8,000 full-time researchers, a $1 billion a year R & D budget, and major nuclear energy, mining technologies, coal-gas liquefaction, and weapons programs has the strongest S & T capabilities in all of Africa, including Egypt. Its achilles heel is that almost all of its researchers are drawn from the 4.6 million whites, ten percent of the total population.[25] South Africa must open its major universities and research centers to the majority of its people - African, Coloured, and Indian - or else risk that its qualified manpower pool stagnates or even declines. This entails massive investments in science education at all levels, a commitment the South African government acknowledges but has yet to make.

Many petroleum exporting countries are attempting to build domestic capabilities through major investments in petrochemicals, natural gas liquefaction, and other petroleum value added technologies. This seemingly sensible approach depends on importing state of the art technologies, relying on highly uncertain future downstream markets, and neglecting research on domestic agriculture, appropriate technologies, small-scale industry, and much else. Scholarships are available to send thousands of students abroad and local institutions are built to foreign scales and models. Even entire research institutes and science and technology policies are imported as in Kuwait and Saudi Arabia. There is little or no concern for learning by doing, maintenance or other informal skills and adaptation is often discouraged, especially with foreign exchange available for the latest imports.

While it may be possible to construct, operate and market petrochemical industries in these countries to augment exports of crude, there is little or no inter-industry linkage. Faculties of geology and petroleum engineering and entire research institutes as in Algeria and Saudi Arabia may be relatively strong while general educational standards are lax. S & T capabilities in the petroleum and petrochemical sector are not readily generalizable across the society, especially since petroleum is so capital-intensive. The large-scale import of contract researchers, often from India and Pakistan, makes locals into administrators and denies them research experience. As long as foreign exchange is available countries may continue to purchase the latest technologies without contributing to indigenous capabilities.

It is hardship rather than affluence that breeds local innovation. It is Venezuela, Trinidad and Tobago and Algeria with their limited quality crude oil reserves that have made the most serious efforts to build nonpetroleum capabilities but with very limited success. Venezuela's Institute for Scientific Investigation has significant basic and applied research capabilities but there are not enough Venezuelan researchers and national training is totally inadequate. Trinidad has put some effort into agriculture and marine biology but also cannot replace its manpower. Algeria has innovated with public sector research and training institutions for petroleum and energy but is plagued by bureaucracy and lack of user ties. Iran, Iraq, Kuwait, and Saudi Arabia have unique situations while sharing an inability to adapt transferred technologies.[26]

Several small countries with impressive human resources are longshots in the science and technology sweepstakes. The combination of generalized literacy, high educational motivation, and a pool of research manpower gives them a chance in spite of the lack of many of the other criteria. The major obstacle is brain-drain which already occurs on a significant scale in each of these countries. However by selecting one or two applied research sectors and networking closely with local firms and multinational investors these countries can find high-technology export niches. These are enterprises with ties to local universities or research centers, relying on local maintenance and spare parts and different from the offshore industries. Each of the longshot countries has invested substantial public and private savings in education, the down-payment on a domestic capability.

There are also several countries as different as Chile and Cuba, Egypt and Nigeria, Indonesia, the Philippines, and Vietnam, which have apparently failed to acquire domestic capabilities. One must say apparent because they continue to try and setbacks need not be failures. However certain patterns can be identified. Cuba and Vietnam have put in place highly centralized science and technology policies based on the Soviet model which isolate basic from applied research, and discourage shop-floor adaptation. The results have been extremely poor but the regimes continue to cling to inappropriate models.

Chile under a right-wing military regime has gone in the other direction turning universities and research institutes into applied research shop-floors. Basic and good applied research has withered and ties with users battered by macroeconomic problems have not developed.

Egypt, Indonesia, Nigeria and the Philippines appear to have so expanded university enrollment and lowered standards that it has become extremely difficult to replace the small number of qualified researchers, a problem aggravated by brain-drain. Inter-industry linkages are frustrated and local adaptation severely limited, especially in Egypt where state enterprises predominate.[27] Formal institutions do not work or work badly and informal learning is frustrated with the result that domestic capabilities are actually declining. Nigeria is drifting into a comparable situation with low academic standards, inability to replace scarce manpower, and import-substitution industries that prefer to rely on technology transfers.

Another 130 developing countries do not appear in these ratings except as "Others." This means that using our criteria their present science and technology capabilities are negligible. This is not entirely a matter of size since Hong Kong, Singapore, and several other rated countries have present populations of five million or less. However many of the other countries have a handful of indigenous researchers. The starkness of their situations is revealed in Tables 1.4 and 1.5 on World Bank agricultural research expenditures in 1975, and research scientists per major crop by developing country. Many developing countries lack even the minimum

TABLE 1.4
Expenditures on Agricultural Research by Region and Income Group, 1975

Region	Number of countries	Agricultural GDP as percent of total GDP	Research expenditures as percent of agricultural GDP	Research expenditures per capita of agricultural population	Number of agricultural research scientists	Number of research scientists per million of agricultural population	Research scientists per million US dollars of agricultural GDP	Research expenditure per scientist per man year	Regional distribution of research expenditure
				(US$)				('000 US$)	(%)
Low Income[1]									
Asia	7	39	0.15	0.14	8,950	14	0.16	10	19
N. Africa/Middle East	3	35	0.29	0.39	970	20	0.15	20	4
West Africa	12	35	0.65	1.14	2,960	42	0.24	27	17
East Africa	10	35	0.38	0.26	790	8	0.12	32	5
Latin America	2	24	0.04	0.06	90	14	0.12	4	Negligible
Total	34	37	0.26	0.26	13,750	16	0.17	15	46
Middle Income[1]									
N. Africa/Middle East	5	20	0.46	1.01	590	35	0.16	29	4
West Africa	3	39	0.32	1.15	510	47	0.13	24	3
East Africa	1	13	1.47	1.33	80	22	0.25	60	1
Latin America	14	13	0.42	1.36	5,440	53	0.16	26	30
Total	23	14	0.42	1.30	6,610	50	0.16	26	38
High Income[1]									
Asia	2	26	0.18	0.61	1,260	58	0.17	10	3
N. Africa/Middle East	3	9	0.26	0.97	560	29	0.08	33	4
Latin America	3	10	0.52	6.30	1,140	169	0.14	37	9
Total	8	12	0.33	1.57	2,960	62	0.13	25	16
All Income									
Asia	9	37	0.16	0.15	10,210	16	0.16	10	22
N. Africa/Middle East	11	15	0.31	0.64	2,120	25	0.12	26	12
West Africa	15	36	0.57	1.14	3,460	43	0.22	26	20
East Africa	11	32	0.43	0.30	860	9	0.13	34	6
Latin America	19	12	0.42	1.51	6,660	56	0.15	27	40
TOTAL	65	21	0.31	0.44	23,320	23	0.16	20	100

[1]Low income is defined as GNP per capita, in terms of 1975 dollars, of no more than $250. Middle income is GNP per capita of $251 to $1,000. These are not official definitions of the World Bank.

Sources: Adapted from Peter Oram, "Current and Projected Agricultural Research Expenditures and Staff in Developing Countries," International Food Policy Research Institute Working Paper 30 (Washington: IFPRI, November 1978). Underlying references include: *Current Agricultural Research Information System* (CARIS), 1973, FAO, Rome; James Boyce and Robert H. Evenson, *Agricultural Research and Extension Programs*, New York, 1975; *List of Research Workers in the Agricultural Sciences*, Commonwealth Agricultural Bureau, London, 1975; International Financial Statistics, IMF, Washington, May 1978; *Yearbook of National Accounts Statistics*, 1976, UN, New York, 1977; *World Tables, 1976*, World Bank 1976; FAO *Production Yearbook*, vol. 30 FAO, Rome, 1976; *Statistical Yearbook, 1974*, UNESCO, Paris, 1975.

TABLE 1.5
Number of Research Scientists per Major Crop by Developing Country[1]

No single crop over 100,000 ha.	Number of scientists per major crop			
	0-10	11-20	21-50	More than 50
Brunei	Afghanistan	Botswana	Indonesia	Bangladesh
Cyprus	Chad	Benin	Nepal	India
Hongkong	Ethiopia	Cameroon	Sri Lanka	Republic of Korea
Kuwait	The Gambia	Liberia	Iraq	Malaysia
Lebanon	Mali	Madagascar	Morocco	Philippines
Maldives	Mauritania	Zaire	Pakistan	Thailand
Oman	Niger	Angola	Nigeria	Egypt
People's Democratic Republic of Yemen	Somalia	Tanzania	Sudan	Jordan
Cape Verde	Upper Volta		People's Republic of the Congo	Iran
Comoros	Burundi		Ghana	Libya
Gabon	Central African Republic		Ivory Coast	Syrian Arab Republic
Guinea-Bissau	Rwanda		Sierra Leone	Tunisia
Mauritius	Togo		Malawi	Turkey
Reunion	Uganda		El Salvador	Senegal
Swaziland	Lesotho		Honduras	Kenya
French Guyana	Mozambique		Nicaragua	Zambia
Barbados	Haiti		Bolivia	Mexico
Belize			Ecuador	Argentina
Costa Rica			Paraguay	Brazil
Dominica			Peru	Chile
Grenada			Venezuela	Colombia
Guadeloupe			Uruguay	
Jamaica				
Montserrat				
Panama				
St. Lucia				
St. Vincent				
Trinidad and Tobago				

[1]A major crop is defined as one with a country area exceeding 100,000 ha. This is derived from P.A. Oram's paper, "Training Requirements for Research and Its Application—An Overview" (Washington: CGIAR, 1977). Oram suggests that a major crop commodity research program would require about 44 research scientists.

Sources: *Report of the Task Force on International Assistance for Strengthening National Agricultural Research* (Washington: CGIAR, 1978), p. 5. Estimates of the number of scientists are derived from various sources. Crop area is provided by the *FAO Production Yearbook*, 1976.

present capabilities to do research on their major food or export crops. Yet without such a local capability they cannot be expected to effectively transfer agricultural research. The largest country rated "other" is Bangladesh with a population over ninety million.

Implications

What are the implications of these ratings? What do they suggest to us about the shape of the future given the very long lead times involved in the upgrading of national capabilities?

Globally these ratings indicate that the North-South division of the world economy is likely to persist and to aggravate. The disproportionate share of world output, trade, industrial production, and other scarce resources held by the North will continue. The OPEC countries will not be able to convert their transfer of financial resources and technology into a permanent redistribution of world wealth although their own situations will have improved.

World trade is likely to become increasingly R & D intensive with only a handful of developing countries capable of taking part. Most developing countries will have to rely on labor-intensive technologies to earn needed exports although a few will achieve high-technology export niches.

The most valuable international resources will be research manpower rather than petroleum, or any other commodity. The global distribution of R & D manpower and its productivity will become the single most important variable in world trade. The implications for international migration suggest an intensified global pursuit for scarce research talent.

The ratings have important implications for the US. The evolution of developing country capabilities makes them the present and probable future foremost markets for US technology transfers. Figures 1.2 and 1.3 on US receipts from royalties and fees and the US trade balance for R & D intensive manufactured products illustrate the striking emergence of these countries as major markets. As John Sewall of the Overseas Development Institute has argued the middle-income developing countries represent America's most promising trade partners.

The ratings also reinforce the view that the Pacific Basin and Rim countries will supplant Western Europe in US economic relations. It is here that science and technology

FIGURE 1.2
U.S. Receipts of Royalties and Fees from Selected Nations

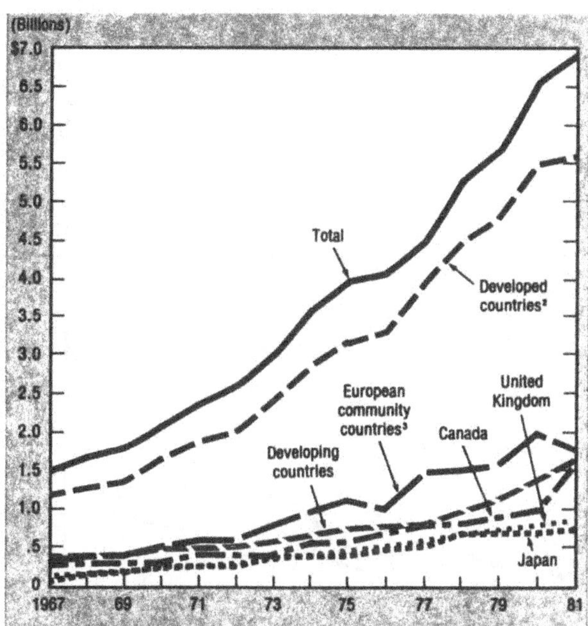

[1]Represents net receipts of payments by U.S. firms from both their foreign affiliates and unaffiliated organizations for the use of intangible property such as patents, techniques, processes, formulas, designs, trademarks, copyrights, franchises, manufacturing rights, management, etc. Excludes firm rentals which are included in the royalties and fees data in the international transaction tables of the *Survey of Current Business*.

[2]Developed countries included here are Western Europe, Canada, Japan, Australia, New Zealand, and the Republic of South Africa.

[3]Original six members only.

Science Indicators—1982

FIGURE 1.3
U.S. Trade Balance[1] with Selected Nations for R&D-Intensive Manufactured Products

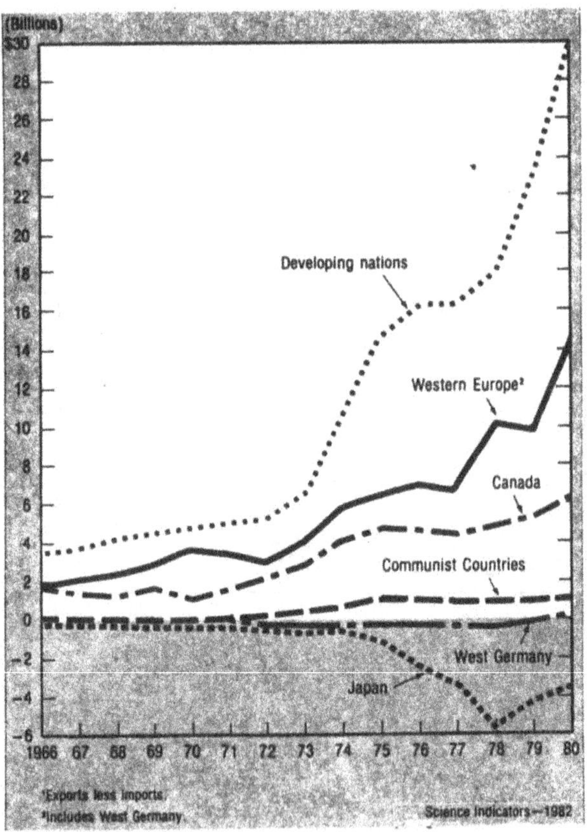

capabilities are most rapidly evolving and where ability to adapt transferred technology is growing. The US has several advantages in competing with Japan as a supplier of R & D intensive goods to these countries, as well as providing a market for their technology adaptive exports.

The ratings also suggest possible changes in regional balances of economic power. The failure of any Middle Eastern state except Israel to achieve S & T capabilities will perpetuate interstate rivalries. Egypt loses its chance to be the S & T center of the Arab world but no one takes its place. South Africa continues to dominate Southern Africa in the absence of significant S & T capabilities anywhere else on the continent. The lack of Middle Eastern and African indigenous capabilities reinforces efforts to involve external powers in regional conflicts. India dominates South Asia economically and militarily with Pakistan nervously seeking external assistance, working hard to improve its own capabilities, but falling further and further behind. The implication is that Pakistan may seek its own nuclear weapons capability as an offset for the much greater overall Indian S & T strength. Japan experiences increasing economic conflicts with regional economic rivals demanding that it open its market to their technology exports. Brazil furthers its present S & T lead over Argentina and Mexico but national import-substitution policies in several countries frustrate its bid for economic domination. Whether or not these projections are valid, science and technology capabilities become key variables in determining regional political and economic balances.

Another implication points to the importance of South-South economic relations. The enormous divergences in S & T capabilities between developing countries facilitate flows of technology adapted to local conditions. Third World multinationals will continue to export to and invest in the markets of countries with lower levels of technology, taking advantage of their own learning by doing, using, and related skills. There will be some substitution of North-South technological dependency with South-South patterns as already exists between Bolivia and Argentina, or Paraguay and Brazil.

The ratings represent a particular point in time in a lengthy process. They suggest that the informal learning skills are more important than the formal criteria of R & D expenditures, manpower, exports, and _in situ_ research. This is because the informal skills are the most deeply embedded in national cultures and the least subject to external transfer. Thus Japan in 1880 might have scored poorly on formal capabilities but would have been seen by astute observers as scoring high on informal learning. The fascinating comparative study by Ranis and Saxonhouse on the nineteenth- and twentieth-century Japanese and Indian cotton textile industries makes this point.[28]

Lessons

The process whereby countries achieve indigenous capabilities is not well understood, as Rosenberg has noted. This study has identified from the experiences over time of a number of countries several important cross-cultural generalizations. These are elements of the process which have appeared sufficiently frequently in time and space to be considered as fundamental.

1. Science education and broad science literacy matter. A widely literate population with access to popular science and technology is an essential asset for informal learning skills. China, Singapore, Taiwan, South Korea and other countries attest to the returns from investment in science education for adults and children. The lack of same is one of the major liabilities of the petroleum dependent states.

2. Elitist advanced education in or out of universities is essential to produce high quality manpower. Where increased enrollment serves to lower standards as in Mexico, Indonesia, Egypt, China during the Cultural Revolution and elsewhere elitist institutions can no longer do their job. India has managed to retain a small number of high quality universities and research institutes in the face of rising enrollment. The education and training of quality researchers requires a hothouse intellectual atmosphere inimical to mass education.

3. Young researchers need apprenticeships to qualified seniors. Where this is absent as in Egypt academic research degrees are of little value. Institutional innovations are essential to permit the master-apprentice system to develop for researchers without rigid age or bureaucratic hierarchies. A related problem is the need for critical masses of researchers to permit team and interdisciplinary work. Isolation is the worst enemy of developing country researchers. Scarce competent researchers should be used as masters for apprentices rather than as administrators or classroom teachers.

4. Continuity of leadership and funding is essential to build research capabilities. The ability of Argentina to

maintain a respectable nuclear research program over twenty years of civil strife and economic disorder is the outstanding example of this lesson. The achievement of research capabilities requires long-term planning and commitment, even for applied research.

5. The extensive and persistent politicization of universities in many developing countries demands nonacademic institutional innovation. National research institutes in Kuwait, South Korea, Venezuela and elsewhere which also offer advanced education and training are one approach. Nonprofit research centers are another widely used device in Brazil and Argentina, especially in the politically sensitive social sciences. The failure to innovate research outside universities has badly hurt Turkey. West Germany and Japan have extensive experience in institutional innovation for high-quality basic and applied research and training outside universities.

6. The networking of users and researchers depends on effective prior networking of users and users, researchers and researchers. The role of national voluntary organizations such as the associations for the advancement of science in Brazil, Mexico, India, and other countries has been underestimated. Researchers need tangible financial and psychological rewards and voluntary groups can help provide the latter. The organization of the scientific community is a priority agenda item in the achieving of indigenous capabilities. The community must have sufficient autonomy to act as a pressure group rather than a government mouthpiece. Brazil and India have set the lead.

7. High priority must go to agricultural research. It has the greatest potential for quick productivity gains that can both raise general welfare and provide surpluses for other R & D. Building agricultural research capabilities is the only short-cut.

8. Technology transfer needs to be recast as a means of achieving domestic capabilities. This involves forswearing import substitution unless it has local learning built-in. It also suggests emphasizing transfers that promote local using and doing rather than elaborate and bureaucratic screening mechanisms. The heart of the matter should be the relationship between provider and client rather than the terms of the transfer per se. Local learning should also be rewarded by access to foreign exchange from technology exports.

9. This study and others argue that the informal skills involved in mastering imported technology are more important

than the formal R & D structure. Public sector enterprises are particularly weak at these informal skills and need drastic management innovation. Since developing country petroleum, transportation and other state-owned and managed enterprises are the largest firms in many countries attention to changes that will improve their informal learning is urgent. One approach would be to ration their access to investment capital and foreign exchange in relation to their ability to export technology.

10. Science and technology planning needs to concentrate more on fostering networks and linkages than on detailed financial and physical planning. Although many countries have established ministries of science and technology or government agencies rarely do these have substantial political or economic clout. Instead they need to serve as facilitators bringing together the scientific community and entrepreneurs, basic and applied researchers, scientists and science educators, while promoting institutional innovation. Top-heavy, centralized bureaucratic science and technology planning is an obstacle in a number of countries.

The road from transferring technology to achieving institutionalized capabilities is long, risky, full of potholes and blind curves, and treacherous. At the end of the road is not a pot of gold but a fresh lease on life for the nation-state as a world actor and a focus for the energies of its people.

NOTES

1. Charles Cooper, Policy Interventions for Technological Innovation in Developing Countries, World Bank Staff Paper 441, Washington, DC, 1982.

2. A. G. Kenwood and A. L. Lougheed, Technological Diffusion and Industrialization Before 1914 (New York: St. Martin's, 1982).

3. Nathan Rosenberg, Inside the Black Box: Technology and Economics (New York: Cambridge, 1982).

4. Louis T. Wells, Jr., Third World Multinationals: The Rise of Foreign Investment from Developing Countries (Cambridge, Mass.: MIT Press, 1983).

5. Rosenberg, op. cit., p. 8.

6. Ibid., p. 271.

7. Ibid., p. 273. Quoted from Henry Rosovsky, "What are the Lessons of Japanese Economic History?" in A. J.

Youngson, ed., Economic Development in the Long Run (New York: St. Martin's, 1972).

8. Rosenberg, p. cit., p. 248.

9. Ibid., p. 275.

10. Jean-Louis Reiffers, Transnational Corporations and Endogenous Development (Paris: UNESCO, 1982).

11. Paulo Bastos Tigre, Technology and Competition in the Brazilian Computer Industry (New York: St. Martin's, 1983).

12. Baldev Raj Nayar, India's Quest for Technological Independence, 2 Vols. (New Delhi: Lancers, 1983). Vol. I, pp. 339-410, "Technology Transfer Policy in the Post-Nehru Era."

13. National Science Foundation, Science Indicators 1982, pp. 25-26, Washington, DC, 1983.

14. Nayar, op. cit., Vol. I. Chapters 4 and 6 on Science Policy.

15. Wells, Jr., op. cit., pp. 81-82.

16. World Bank, Agricultural Research, Sector Policy Paper, Washington, DC, June 1981, pp. 25-26.

17. Ibid., p. 6.

18. Francisco Sagasti and Alberto Araoz, eds., Science and Technology for Development: Planning in the STPI Countries (Ottawa: International Development Research Center, 1979) and Science and Technology for Development: Main Comparative Report of the Science and Technology Policy Instruments Project (Ottawa: IDRC, 1979).

19. Nayar, op. cit., Vol. I, pp. 537-538 summarizes the Indian research and development infrastructure as of 1980.

20. Leo Orleans, ed., Science in Contemporary China (Stanford: Stanford University Press, 1980).

21. Simon Schwartzman, Formacao de Comunidade Cientifica No Brasil (Sao Paulo: FINEP, 1979).

22. Ibelis Velasco, "Algunos Hechos y Muchas Impresiones Sobre La Ciencia y La Tecnologia en Argentina," Interciencia, Part I, Vol. 8, No. 3, May-June 1983, pp. 166-172; Part II, Vol. 8, No. 4, July-Aug. 1983, pp. 224-232.

23. Miguel Wionczek, "Science and Technology Planning in Mexico and its Relevance to other Developing Countries," in Francisco Sagasti and Alberto Araoz, eds., op. cit., pp. 90-108. Ibelis Velasco, "Algunos Hechos y Muchas Impresiones Sobre Giencia y Tecnologia en Mexico," Interciencia, Part 1, Vol. 6, No. 6, Nov.-Dec. 1981, pp. 402-408; Part II, Vol. 7, No. 1, Jan.-Feb. 1982, pp. 37-44.

24. Ziauddin Sardar, Science and Technology in the Middle East (New York: Longman, 1982). A. B. Zahlan, Science

and Science Policy in the Arab World (London: Croom Helm, 1980).

25. Aaron Segal, U. S.-South Africa Relations: The Technological Factor (Bloomington: African Studies Program, Indiana University, 1984).

26. Sardar, op. cit. A. B. Zahlan, ed., Technology Transfer and Change in the Arab World (London: Pergamon, 1978).

27. Clement Henry Moore, Images of Development: Egyptian Engineers in Search of Industry (Cambridge, Mass.: MIT Press, 1980).

28. Gustav Ranis and Gary Saxonhouse, "International and Domestic Determinants of Technology Choice by the Less Developed Countries," pp. 7-29, in Barbara Lucas and Stephen Freedman, eds., Technology Choice and Change in Developing Countries: Internal and External Constraints (Dublin: Tycooly, 1983).

2

Latin America: Development with Siesta

Aaron Segal

"Either Latin America dominates technology or through other countries technology will dominate Latin America." These words of the Latin American scientist and science and technology policy advocate Jorge Sabato who died in 1983 have become deeply embedded in the Latin American and Caribbean psyche. Nearly every country during the last two decades has initiated a national effort to substantially improve its scientific and technological capabilities. Jorge Sabato and others who launched these national and subregional and regional campaigns in the 1960s have been extraordinarily influential within the scientific and technological communities themselves and with many governments. One aspect of this influence is the desire to construct an original model of the organization of science and technology for development rather than to copy what prevails elsewhere. Sabato insisted that "We want development, but with siesta."[1]

What then is the current status of efforts to build indigenous science and technology capabilities in Latin America and the Caribbean? How and to what extent have these capabilities been mobilized for economic development and other goals? The assessment of capabilities is partly indicated in national and regional expenditures on research and development (R & D) although these data are often less than comparable between countries and not yet collected on a uniform basis. (Table 1.2) Data are also available on a non-uniform basis on R & D by countries. (Table 2.1) The formation of human resources with science and technology is reflected in a rapid growth in enrollment, further lowering of academic standards, and a remarkable expansion of private universities.

Assessing national and regional technological balances

TABLE 2.1
Latin America: Total Research and Development Costs and as a Percentage of Gross National Product (GNP)

Country	Year	U.S. Dollars (thousands)	% of GNP
Large Countries			
Argentina	1978	245,386	0.39
Brazil	1978	1,150,028[a]	0.61
Mexico	1980	371,739	0.24[b]
Andean Countries			
Colombia	1978	20,600	0.11
Chile	1979	65,652	0.33
Ecuador	1979	11,627	0.13
Peru	1976	48,111	0.36
Venezuela	1977	201,616	0.56
Other South American Countries			
Paraguay	1971	1,328	0.20
Uruguay	1972	3,300	0.15
Central America			
Costa Rica	1981	5,186	0.17[e]
El Salvador	1974	4,760	0.31
Guatemala	1978	13,504[e]	0.22[e]
Honduras	1971	1,481[c]	0.20
Nicaragua	1971	1,121[c]	0.14
Panama	1975	3,296	0.17
Caribbean			
Cuba	1978	112,270	n.d.
Jamaica	1973	6,820	0.36
Dominican Republic	1972	1,561	0.08
Trinidad and Tobago	1970	2,586	0.32

[e]Estimates
[a]Does not include the private sector
[b]As a percentage of gross domestic product
[c]Figures refer to only two research centers

Sources: Jan Annerstedt, A Survey of World Research and Development Efforts, Institute of Economics and Planning, Roskilde University, Denmark: UNESCO, La politica cientifica y tecnologia en America Latina y el Caribe-4, Estudios y Documentos de Politica Cientifica, No. 42, Anuario Estadistica 1980, Estadisticas sobre el personal cientifico y tecnico y los gastos destinados a actividades de investigacion y desarrollo experimental en America Latina y el Caribe, 1981, International Monetary Fund, International Financial Statistics 1981. Some data for Brazil, Costa Rica, and Mexico were obtained from unpublished sources. Francisco R. Sagasti, "Ciencia y tecnologia en America Latina," Comercio Exterior, Mexico, December 1984, p. 1172. Vol. 34, No. 12, published with permission.

requires access to other kinds of data. Table 2.2 provides estimates for several countries of the recent costs of imported technology as reflected in copyrights, patents, royalties, licenses, and trademarks. Brazil and to a lesser extent Argentina and Mexico have become significant exporters of technology in various forms. The export of manufactured goods can also be used as a rough measure of technological capabilities although the degree of indigenous adaptation, design, and engineering contribution to the final product varies.[2]

Another concept of technological balance refers to Latin American participation in new science-driven applied research technologies such as microelectronics, biotechnology, remote sensing, marine sciences, and other promising fields. Here "state of the art" assessments rely on publications in leading world scientific journals, citation counts of the use of articles, and other quantitative and qualitative data to examine to what extent Latin American scientists are participating in this work and at what levels.[3] Even where basic research has moved into applied research and development stages and proprietary data is emerging, publications are still a useful indicator of whether local scientists are keeping up with major research trends.

While there is then no simple answer to the status of science and technology in thirty countries that differ enormously in size and resources, it is possible to combine several criteria to provide an over-all assessment. This was performed at the second meeting of the Council of the Latin American Academies of Science (ACAL) held in Chile in April 1984.[4] Organized in 1983, the council represents national academies of science rather than governments. Its report lauds the creation within thirty years of a Latin American scientific community reflected in regular meetings, journals, publications, original research, and the training of future scientists. It goes on to note that with 10 percent of the world's population Latin America has only one percent of the world's published research and only a handful of internationally known scientists. (Only one Latin American is in the world's 1,000 most cited scientists.) Moreover, 92 percent of Latin American scientific published output comes from only five countries (Argentina, Brazil, Mexico, Chile, and Venezuela).

The ACAL Report finds that proportionately Latin American countries are spending 5-10 times less per capita on R & D than the post-industrial societies. No Latin American state has come close to investing one percent of

TABLE 2.2
Latin America: Payments for the Import of Technology by Types

Country	Direct Foreign Investment (1980) Payment	%	Imports of Capital Goods (1979) Payment	%	Royalty and Rights Payments Payment	%	Total Payment	%
Large Countries								
Argentina	740.6	24.5	2,175.8	72.1	101.0[a]	3.3	3,017.4	100
Brazil	1,568.3	28.4	3,444.9	62.5	500.0[b]	9.1	5,513.2	100
Mexico	1,852.1	22.9	5,781.1	71.4	462.7[c]	5.7	8,095.9	100
Andean Countries								
Bolivia	41.5	11.6	313.7	87.9	1.8[d]	0.5	357.0	100
Colombia	233.0	16.2	1,195.3	83.3	6.32[d]	0.4	1,434.6	100
Peru	26.9	3.4	749.2	95.6	7.48[c]	1.0	783.6	100
Venezuela	54.7	11.0	4,334.8	87.0	101.0[d]	2.0	4,982.8	100
Central American and the Caribbean								
Costa Rica	12.2	4.1	275.8	93.7	6.5[e]	2.2	294.5	100
Guatemala	110.0	23.0	360.3	74.4	12.7[e]	2.6	484.0	100
Trinidad	216.5	41.8	283.2	54.7	18.0[f]	3.5	517.7	100

[a]1974 [b]1977 [c]1980 [d]1979 [e]1976 [f]1975

Sources: Inter-American Development Bank, Economic and Social Progress in Latin America.1982 Report; BIB-CEPAL, Estudio Economico de America Latina 1980; Sintesis estadistica de America Latina 1960-1980, y F. Sagasti y C. Paredes; La situacion de la ciencia y la technologia en America Latin, el Caribe, (GRADE, Lima, March 1982). Francisco Sagasti, "Ciencia y tecnologia en America Latina," Comercio Exterior, Mexico, December 1984, p. 1172. Vol. 34, No. 12, Published with permission.

its gross national product (GNP) in R & D, compared to over 2 percent annually in the United States, Japan, and elsewhere. Thus the absolute gap in science and technology is growing constantly, with the U.S. public and private sectors spending more on research in one year than all of Latin America in several decades.

The report cites a number of major obstacles to research. These include political and economic instability, which results in: a continuing brain-drain of researchers; non-participation of scientists in decisions concerning research projects; rigid systems of higher education which emphasize "professional" rather than research degrees; many postgraduate courses taught by instructors who lack research experience; isolation and lack of contact and exchange between Latin American scientists; separation of scientific communities from economically productive sectors; and "insufficient community and government understanding of the role of science in development." These factors serve to perpetuate the "backwardness and dependency of our countries."[5]

It is easier to diagnose what ails science and technology in Latin America than to prescribe remedies. The report wants respect, merit, and adequate working conditions for researchers; sustained government support for public- and private-research organizations; the inclusion of qualified researchers in policy-making for science and technology; the establishment of careers in research; duty-free import of research equipment and scientific publications; and scholarships for high-quality research-oriented masters and doctoral degrees.

Recognition that improvements in science and technology require a sub-regional or regional context has been present for many years. Not even Brazil as the most important science and technology power in the region can afford at the national level to pursue more than a limited number of research areas. As science-driven team-conducted interdisciplinary research prevails, various forms of regional cooperation become increasingly attractive. The Council of Latin American Academies of Science proposes systematic exchanges of researchers through designated centers of excellence and postgraduate courses, workshops on specific research topics, and working groups to explore collaboration with production sectors.

The profound economic recession of 1979-1986, the deep budget cuts in many countries, and the staggering external debt burden have all served to strengthen the arguments for indigenous science and technology capabilities that began in the 1960s. There is widespread disillusionment with high-cost, inefficient import-substitution that relies heavily on imported capital-intensive technologies. There is an urgent need to increase the exports of manufactured goods in order to service external debts and to earn foreign exchange. There is an awareness that the regional and subregional economic integration schemes of the 1960s and 1970s, such as the Latin American Free Trade Area (LAFTA), the Central American Common Market (CACM), the Andean Pact, and the Caribbean Community (CARICOM), have neglected the technological dimension and broken down in squabbles over static trade rivalries.[6] There is a growing sense that a failure to participate in new research areas such as biotechnology and microelectronics will condemn Latin America to generations of new expensive imports without the local ability to adapt or to improvise. Perhaps most important, neither the laissez-faire monetarist economists nor the Marxist ideologists have credible alternative ideas about development. Science and technology have won a place in the battle of ideas about shaping the future even if their hold on the present remains shaky in some countries.

Thus the National Science and Technology Councils formed in the 1960s and 1970s in nearly every country in Latin America (known usually by the Spanish acronym CONACYT) have survived the recession intact as have major research centers.[7] Science and technology have achieved in the major countries a constituency consisting of national scientific communities including the national associations for the promotion of science which exist in seven countries; leading politicians and military who support indigenous science and technology; some university and local private businesses; and even multinational corporation support. These national constituencies can rely on external support and some funding from multiple multilateral and bilateral sources and in particular the World Bank, the Inter-American Development Bank (IDB), UN agencies such as the World Health Organization, the Economic Commission for Latin America, and the Organization of American States (OAS). Increasingly visible in support of science and technology efforts are Latin American regional organizations such as the Latin American Economic System (SELA) with its regional technical information network project, the Latin American Energy Development Organization (OLADE), and dozens of recently formed associations of Latin American chemists, historians of science, physicists, marine biologists, biochemists, and others. A February 1984 meeting in Mexico even established a Latin American Association for Science and Technology

Policy (ALPCYT).[8]

Four forces continue to drive Latin American and Caribbean science and technology in 1987 as they have for more than twenty years. The first and foremost is nationalism based on the assumption that a modern nation-state without a significant indigenous science and technology capability is both dependent and vulnerable. This is the force behind the creation of a Brazilian national computer industry, the Argentine ability to sustain nuclear reactor research during twenty years of economic and political chaos, Mexican efforts to organize marine-biology centers, and other national efforts.[9]

A second force consists of the desire to be able to competitively screen and select technology imports based on national needs. Most Latin American governments during the 1970s passed laws to regulate the import of technology. These laws require importers to "unbundle" imported technological packages and to negotiate competitive terms for each item.[10] These government efforts to regulate technology imports have produced much frustration and few clear benefits but they continue. Regulation of technology transfers has become part of the effort to reduce imports and to manage external debts.

During the worst of the recession Brazil, Mexico, and other countries were able to sharply reduce imports of consumer goods. However cutting back on technology imports came with a high price in losses in productivity, inability to produce capital goods, and, worst of all, losses of exports because of inability to guarantee quality control. Hence technology transfer regulation has become intimately linked to the promotion of domestic science and technology capabilities.

A third force has been disenchantment with industrialization based on import substitution policies.[11] The limited size of national markets, the high costs and subsidies of protectionism, the overreliance on unmodified imported technology, the need to earn foreign exchange and to diversify exports, and the knowledge that the fastest growing most recession-proof developing economies in the world have been technology exporters such as Taiwan, South Korea, Singapore, and Hong Kong has produced a new mind-set. A national science and technology infrastructure is considered a prerequisite for export-industrialization in countries as different as Argentina and Mexico, Brazil and Colombia, and Chile and Venezuela. Brazil in the 1980s was generating more than one-third of its total export earnings from domestic or adapted foreign technologies, selling computers, telecommunications systems, passenger jets, and processed orange juice to the rest of the world. Colombia, Mexico, Chile, Argentina, and other countries were steadily increasing the quantity and quality of their manufactured exports incorporating domestic technologies, although lagging far behind Brazil. Import substitution protectionist policies are giving way in country after country because of the belief that their effects are not good for local science and technology.

The fourth driving factor is the belief that science and technology are major components of economic growth, whether in developed or developing economies. Jose Pastore has calculated that over 25 percent of all Brazilian economic growth over a fifty year period has been due to science and technology.[12] Diffusion of the hybrid seeds developed by international research based in Mexico has contributed significantly to increased maize yields throughout Latin America. Major advances in controlling epidemic diseases such as smallpox, cholera, and polio have been achieved through applied research, demonstration, and diffusion. Similarly the lack of *in situ* local research or government unwillingness to pay attention to research findings has contributed to ecological and economic disasters in the Brazilian Amazon, to the devastation of Peruvian fishing catches, and to costly abandoned commercial nuclear reactor projects in Brazil and Mexico.[13]

National scientific communities maintain that major development problems involving natural resources and *sui generis* climactic or environmental conditions can be tackled only by local research. Tropical agriculture, marine biology, hydrology, geology, applied biotechnology, and other subjects cannot acquire foreign off-the-shelf technologies which will fit *in situ* research needs. There is no substitute for a local R & D effort, whether dealing with Andean and Central American earthquakes or Caribbean coral reefs. Multilateral and bilateral programs of research cooperation are generally organized around collaboration between foreign and national scientists as in U.S., Soviet, French, Chinese, Israeli, Canadian, British, Japanese, and other programs.

Just as the driving forces are similar in many of the countries so are the fundamental problems encountered.[14] The most persistent and critical frustration is the lack of linkages between university researchers, government centers, and the private sector. The "triangle of linkages" depicted by the late Jorge Sabato has not been realized. Instead university researchers often concentrate on basic science, publish abroad, and emulate work done in the more advanced societies. Government research institutes are often enfeebled by underfunding, political instability, and high personnel turnover. Their work is often highly applied but without clear clients or users. Local private firms usually prefer to import technology rather than invest in research capabilities and distrust the work done at local universities and government institutes. Multinational firms are generally branch plants with built-in incentives to use parent-firm imported technology and engineering services. Latin American state-owned enterprises have no clear commitment to local technology, suffer from erratic management, unstable subsidies, and limited interest in achieving in-house research capabilities.

The national science and technology councils created during the 1960s and 1970s have had a hard time coordinating policies and facilitating linkages. Most are weak institutionally, vaguely attached to a government ministry or floating in the president's office. Most have been unable to fund more than 5-10 percent of total national R & D and thus have lacked the financial clout to bring together university researchers and possible private sector or other clients. Instead the councils have concentrated on planning, manpower surveys, coordination of international cooperation, organizing conferences and publications, and awarding scholarships for foreign study. Only in Brazil where the National Council of Scientific Development and Technology (CNPQ) commands hundreds of millions of dollars of research funding, Venezuela where the scientific community has participated extensively in council plans, and Costa Rica where the council has brought university researchers together with clients, have linkages become ongoing. For instance in Brazil the CNPQ has funded university centers of excellence in fields such as solid-state physics and aerospace and these centers have worked closely with the private sector and state enterprises.[15]

The irony is that the need for science and technology is widely recognized but the demand for local R & D is uneven and sporadic. Efforts to bring together researchers and users have run into problems. In 1970 Peru established an Industrial Technology Institute that received a 2 percent tax on private firms which it could use to fund research with them, through universities, etc. Controversial and subject to differing evaluations, this scheme of forced linkages was dropped by the Belaunde Terry Government.[16] Mexico's CONACYT, after producing two five-year national plans for science and technology which remained largely formal, unimplemented documents was subject to a major revision in 1984 intended to strengthen its ties to the private sector. The impressively funded and highly productive Venezuelan Institute for Scientific Investigation (IVIC) established in the early 1980s its own autonomous engineering, applied research, and consulting services in order to promote its weak links to the private sector.[17] Other countries such as the Dominican Republic have used external and local funds to establish nonprofit industrial technology centers which provide applied research, consulting, and design for a fee to local users.

One of the principal factors explaining the lack of linkages is the state of higher education in the region.[18] Since 1960 there has been a tenfold increase in undergraduate enrollment, a proliferation of new universities, and the rapid expansion of graduate programs with an estimated 68,000 students in 1980, mostly in Argentina, Brazil, Mexico, and Cuba. These changes have occurred with little or no provision in most universities for funding of research facilities or full-time researchers. The rapid expansion of higher education has perpetuated the bias toward professional degrees, reliance on part-time faculty, and nonprovision for research in the curriculum. Military intervention in a number of countries such as Argentina has worsened matters by forcing researchers to emigrate and breaking up research teams. Although there are pockets of excellent research at certain public and private universities throughout the region these understandably survive by not having links within their societies and by focusing on basic research. Higher education in Latin America and the Caribbean with its emphasis on numbers and its extensive politicization does not provide the environment that sustained scientific research requires.[19]

Another response to the producer-user gap for science and technology has been the evolution of extensive shop-floor R & D. Argentine economist Jorge Katz directed a series of UN Economic Commission for Latin America and other agency-funded studies of applied research performed by specific locally owned companies and industrial sectors in Argentina, Mexico, Colombia, Brazil, and elsewhere.[20] This work has been corroborated by a number of other micro-industry studies by researchers in several countries. In essence, faced with shortages of foreign exchange, slow and unresponsive government bureaucracies, worn-out spare parts, ill-fitting imported technologies, the desire to export to earn foreign exchange, and other characteristics, firms have learned how to adapt. This learning has taken the form of reverse engineering to take technologies apart and to improve usage, adaptation to reduce imports, redesign, and other shop-floor improvisations. The companies are generally locally owned, private, highly competitive, but having no formal research departments, budgets, or staff. A few state firms have also improvised as have multinationals when allowed to do so by head offices.

The extent of shop-floor learning seems to be less than that documented by studies in the Far East but by no means inconsequential. A few efforts in Colombia and Mexico by schools of engineering to work with shop-floor innovators have brought limited results. The evidence suggests that entrepreneurship and innovation are largely to be found within the firms rather than in the overcrowded universities or the stodgy government research centers. There is a much stronger coorelation between the presence of shop-floor learning and the desire to export than any formal or informal linkages with other R & D producing sectors.

The failure to establish linkages has also resulted in many countries in the movement of institutionalized research into nonprofit centers and institutes away from the universities. The list of such research centers and their activities is long and impressive.[21] It includes the Colegio de Mexico with its social science programs in Mexico City, the Di Tella and Bariloche Foundations in Argentina working in the physical and social sciences, Venezuela's IVIC with plentiful full-time researchers, the research and training institute operated by PEMEX for petroleum and petrochemicals in Mexico, and many others. The Inter-American Development Bank and the World Bank recognized this trend in the 1980s by shifting science and technology loans away from universities toward autonomous research centers. This trend is particularly marked in agriculture where Latin America has three internationally funded centers (wheat and maize in Mexico, tropical agriculture in Colombia, and potatoes in Peru), plus a major regional center in Costa Rica while national and especially university R & D for agriculture stagnates. (Table 2.3 and 2.4)

It remains to be seen whether these autonomous research centers can establish effective linkages with users, but their nonpolitical character, relatively stable funding, and provision of full-time research positions makes them eminently more attractive to scientists than university teaching. More and more these centers are taking on the responsibility of providing advanced degrees with research apprenticeships in contrast to the low-quality nonresearch

TABLE 2.3
Human Resources (Professional Personnel) in Agricultural Research in Latin Amercia and the Caribbean, from 1960 to 1980 (Selected Years)*

SUBREGION[1]	1960	1965	1970	1974	1980
Southern Zone (excluding Brazil)	365[2]	816	1,045[3]	1,196[4]	1,364
Brazil	200[5]	500[6]	764	2,000	2,935
Andean Zone	387[7]	643	1,294	1,694	1,843[8]
Panama and Central America (excluding Mexico)	144[9]	305[10]	283[11]	333[12]	383
Mexico	190[13]	279[14]	551	1,000	1,079
Caribbean (excluding the Dominican Republic)	64[15]	96	157[16]	228[17]	198[18]
Dominican Republic	3[19]	5	12[20]	35[21]	99
Latin America and the Caribbean (total)	1,353	2,644	4,106	6,486	7,901

*Preliminary information, still being analyzed (Trigo and Pineiro, 1981: Appendix 2).

[1]Southern Zone includes Argentina, Uruguay, Paraguay, and Chile. Andean Zone includes Bolivia, Peru, Ecuador, Colombia and Venezuela. Central America includes Costa Rica,

(Continued)

TABLE 2.3 (Continuation)

Nicaragua, Honduras, El Salvador and Guatemala. Caribbean includes Guyana, Suriname, Jamaica, Haiti, Barbardos, Grenada, Trinidad and Tobago.

[2] Information for Argentia, Chile and Paraguay is from 1959.

[3] Information for Paraguay is from 1971.

[4] Information for Chile is from 1973, for Paraguay it was estimated at 37.

[5] Information is for 1959.

[6] Information is for 1967.

[7] Information for Bolivia, Ecuador and Peru is from 1959.

[8] Information for Colombia is from 1979.

[9] Information for Honduras and Nicaragua is from 1959; for Guatemala it was estimated at 20.

[10] Information for El Salvador and Guatemala is from 1966.

[11] Information for Honduras, Nicaragua and Panama is from 1971; for Guatemala, from 1972.

[12] Information for El Salvador is from 1973; for Costa Rica and Guatemala it was estmiated at 64 and 58 respectively.

[13] Information is for 1959.

[14] Information is for 1966.

[15] Information is for 1959.

[16] Information is for 1971.

[17] Information for Trinidad and Tobago is from 1971.

[18] Information for Trinidad and Tobago is from 1978.

[19] Information is for 1959.

[20] Information is for 1971.

[21] Estimated.

Reprinted by permission from Martin Pineiro, Eduardo Trigo eds. Technical Change and Social Conflict in Agriculture, Latin American Perspectives, Westview, 1983.

TABLE 2.4

Budgetary Resources Allocated to Agricultural Research in Latin America and the Caribbean, between 1960 and 1980, Selected Years (Constant Value of 1975; Official Money Exchange Rate: National Currency/US dollars, for Year Selected)*

SUBREGION[1]	1960	1965	1970	1974	1980
Southern Zone (excluding Brazil)	31,446[2]	31,298	32,594[3]	44,702[4]	42,559[5]
Brazil	8,280[6]	15,533[7]	24,178[8]	32,879[9]	116,797
Andean Zone	15,631[10]	20,003[11]	43,056[12]	57,393[13]	60,541[14]
Panama and Central America (excluding Mexico)	4,412[15]	4,967[16]	4,904[17]	5,961[18]	10,215
Mexico	4,666[19]	5,218	9,723	14,637[20]	48,357[21]
Caribbean (excluding the Dominican Republic)	1,530[22]	1,530[23]	3,280[24]	2,940[25]	2,128[26]
Dominican Republic	441[27]	496[27]	490[27]	2,278[28]	1,642
Latin America and the Caribbean (total)	66,406	79,045	118,225	160,790	282,239

*Preliminary figures, currently being adjusted (Trigo and Pineiro, 1981: Appendix 1).

(Continued)

TABLE 2.4 (Continuation)

[1]Southern Zone includes Argentina, Uruguay, Paraguay, and Chile. Andean Zone includes Bolivia, Peru, Ecuador, Colombia and Venezuela. Central America includes Costa Rica, Nicaragua, Honduras, El Salvador and Guatemala. Caribbean includes Guyana, Suriname, Jamaica, Haiti, Barbardos, Grenada, Trinidad and Tobago.

[2]Information for Chile is from 1961.

[3]Information for Paraguay is from 1971.

[4]Information for Chile and Uruguay is from 1973; for Paraguay from 1972.

[5]Information for Argentina is from 1979.

[6]Information is from 1962.

[7]Authors' estimate, based on figures supplied by Boyce and Evenson.

[8]Information is from 1972.

[9]Information is from 1973.

[10]Information for Bolivia, Venezuela and Peru is from 1962; for Ecuador from 1965.

[11]Information for Bolivia is from 1962.

[12]Information for Bolivia and Venezuela is from 1972 and 1969 respectively.

[13]Information for Bolivia and Ecuador is from 1973; for Venezuela and Peru from 1976.

[14]Information for Colombia is from 1979.

[15]Information for Nicaragua and Guatemala is from 1962; for Honduras from 1963.

[16]Information for El Salvador is from 1966; for Guatemala from 1962 and Panama from 1961.

[17]Information for Honduras and Nicaragua is from 1965; for Guatemala from 1973; for Panama it was estimated as US$600,000.

[18]Information for El Salvador is from 1973; Honduras from 1976 and Panama from 1975; for Nicaragua it was estimated as US$1,000,000.

[19]Information is for 1962.

[20]Information is for 1972.

[21]Information is for 1979.

[22]Information for Barbados, Jamaica, Suriname, Grenada, Trinidad and Tobago is from 1965; for Guyana it was estimated as US$250,000.

[23]Same information as 1960.
[24]Information for Barbados, Jamaica, Suriname, Grenada, Trinidad and Tobago is from 1972; for Guyana from 1973 and for Haiti from 1976.
[25]Information for Barbados and Haiti is from 1976; for Jamaica, Trinidad and Tobago from 1972.
[26]Information for Haiti is from 1978; for Suriname and Grenada from 1974, and for Guyana from 1978.
[27]Information was estimated on the basis of 10 per cent of the totals for Panama and Central America.
[28]Information is for 1977.

Reprinted by permission from Martin Pineiro, Eduardo Trigo, eds. _Technical Change and Social Conflict in Agriculture, Latin American Perspectives_, Westview, 1983.

graduate programs that are proliferating at many universities.

By 1987 a decade of experimentation with science and technology policies has produced distinct patterns. Brazil is the only country with both a comprehensive infrastructure, domestic linkages, and a coherent operational policy based on national priorities. Current priorities include aerospace, military technologies for export, hydropower, gasohol, telecommunications, computer science, and development of the Amazon region. Rejected priorities, after costly errors, include commercial nuclear reactors via technology transfer, Amazon pulp and paper industries, and several others. The current choice of priorities is highly questionable on both efficiency and equity grounds, especially the infant computer and telecommunications industries.

Argentina has retained a considerable science and technology infrastructure in spite of two generations of brain-drain but has no coherent policy or institutions.[22] Basic research has gravitated to the foundations and a few university centers while applied research is mostly at the shop-floor level. Agricultural yields remain extremely low and basic and applied agricultural research neglected. The Alfonsin Government has taken steps to restore some of the credibility of the very weak National Science and Technology Council but has been unable to redress years of underfunded research.

Mexico poured hundreds of millions of dollars into its science and technology infrastructure prior to the economic crisis of 1981. Although the absolute number of researchers has grown impressively basic research remains isolated, overwhelmingly concentrated in Mexico City, and with much of the scientific community hostile to official goals.[23] The 1984 reorganization proposed to emphasize applied research for the private sector but follow through and modalities are poor. Nor have the technology transfer laws of the 1970s been effective either in spurring local capabilities or in leading to more selective imports. Mexico has made tremendous strides in improving its science and technology infrastructure but has yet to figure out how to mobilize it effectively for development.

Among the middle-level countries Colombia has the best record of science and technology for industrial exports although its infrastructure is spotty and deficient, especially in university research. Venezuela has concentrated on human resources with millions poured into scholarships at home and abroad; it has built an excellent basic-research complex at IVIC, but has a limited record of applied research tied to economic development, especially in agriculture. Chile has experienced two decades of brain-drain, has hung on to several strong academic and nonprofit research pockets but has neither an effective infrastructure nor a policy.[24] Peru has a much weaker infrastructure than Chile, even fewer pockets of excellence, and much government-scientific community hostility after frustrated or abandoned experiments and plans.

Costa Rica stands out among all the smaller Caribbean and Latin American societies for its commitment to and implementation of a science and technology policy.[25] Its priorities are tropical ecology, agriculture, and forestry and marine biology, all consistent with its natural resources and the small size of its scientific community. It has achieved significant researcher participation in official planning and some innovative funding to relate academic research to the needs of Government agencies and parastatal enterprises. Costa Rica has also taken advantage of bilateral and multilateral cooperation in a sophisticated manner.

Other smaller countries such as Jamaica, Trinidad and Tobago, Ecuador, and the Dominican Republic have science and technology national councils, policy documents, and a few institutions, but lag behind Costa Rica in implementation. It is apparent that science and technology policy models for the smaller countries will have to be different.

Cuba stands alone, having opted for the highly centralized Soviet-policy model with the Academy of Sciences operating most major research divorced from productive enterprises.[26] (See Chapter 3.) The formation of scientists is impressive and there have been several significant research successes (interferon production, livestock embryo transplants) but the overall record is unimpressive.

Throughout the region expectations concerning the potential of science and technology are substantially ahead of capabilities. The problems of funding, inadequate information systems, isolation, lack of foreign exchange to purchase periodicals, nonexistent linkages, and poorly defined careers are obstacles to institutionalization, infrastructure, and enhanced capabilities. Improvement can come only incrementally and a decade of large-scale investments can provide returns only if another more costly decade of investments in people, equipment, and institutions is made. Meanwhile new research and technologies generated elswhere flood the market. The region must run as fast as it can in science and technology in order not to get further and further behind.

NOTES

1. Francisco R. Sagasti, "A Man of our Times: Jorge Sabato at One Year of His Death," Interciencia, Vol. 9, No. 6, (Nov.-Dec. 1984), pp. 404-406.

2. Louis T. Wells, Jr., Third World Multinationals (Cambridge: Mass.: MIT Press, 1983), pp. 67-91.

3. Interciencia Association, Biotechnology in the Americas, Washington, DC, 1984.

4. Interciencia, Caracas (Vol. 9, No. 5), Sep.-Oct. 1984, pp. 316-327.

5. Ibid, p. 316.

6. Francisco R. Sagasti, Ciencia, tecnologia y desarrollo latinoamericano (Mexico: Fondo de Cultura, 1981), pp. 233-242.

7. Francisco R. Sagasti, "Ciencia y tecnologia en America Latina," Comercio Exterior, Mexico, December 1984, pp. 1163-1180.

8. Interciencia, Vol. 10, No. 1 (Jan.-Feb. 1985), pp. 40-42.

9. Paulo Bastos Tigre, Technology and Competition in the Brazilian Computer Industry (New York: St. Martin's, 1983).

10. The principles of many of these technology transfer laws were developed in Constantine Vaitsos, Transfer of Technology to Developing Countries Through Private Enterprises (Oxford: Clarendon Press, 1970).

11. The different economic growth paths of export-oriented and import substitution developing countries are discussed in the World Development Report 1984, World Bank, Washington, DC, pp. 23-34.

12. Jose Pastore, "Science and Technology in Brazilian Development," paper presented at the October 1976 U.S. National Academy of Sciences Bicentennial Symposium, Washington, DC, p. 100.

13. Ibelis Velasco, "Some Facts and Many Impressions on Science and Technology in Peru," Interciencia, Vol. 6, No. 5 (Sep.-Oct.1 1981), pp. 345-354, for a discussion of fisheries research and policy in Peru.

14. The Canadian International Development Research Center (IDRC) funded a ten nation comparative study of science and technology policy between 1973 and 1976 including Argentina, Brazil, Colombia, Mexico, Peru, and Venezuela. Results are summarized in Francisco Sagasti, Ciencia y tecnologia y desarrollo latinoamericano, op. cit. pp. 158-195, and are available in English in separate publications from the IDRC, Box 8500, Ottawa, Canada.

15. Simao Mathias, "Evolution of Scientific Research in Brazil," and Milton Vargas, "Evolution of Technology in Brazil," in Interciencia, Vol. 7, No. 6 (Nov.-Dec. 1982), pp. 340-354.

16. Ibelis Velasco, "Some Facts and Many Impressions on Science and Technology in Peru," op. cit. See also Francisco Sagasti, Technology, Planning, and Self-Reliant Development (New York: Praeger, 1979), pp. 117-135.

17. Ibelis Velasco, "Some Facts and Many Impressions on Science and Technology in Venezuela," Part I, Interciencia, Vol. 7, No. 5 (Sep.-Oct. 1982), pp. 301-308; Part II, Interciencia, Vol. 7, No. 6 (Nov.-Dec. 1982), pp. 363-368.

18. Ivan Jaksic, "The Politics of Higher Education in Latin America," Latin American Research Review, Vol. 2, No. 1, 1985, pp. 209-221. Daniel C. Levy, Higher Education and the State in Latin America (Chicago: University of Chicago Press, 1986). Levy traces the remarkable growth of private universities during the last two decades, especially in Brazil. However most private universities lack the resources to conduct research.

19. Mario Bunge, "The Seven Capital Sins of Our University and How to Redeem Them," Interciencia, Vol. 9, No. 1 (Jan.-Feb. 1984), pp. 37-38.

20. Jorge M. Katz, "Technology and Economic Development: An Overview of Research Findings," in Moshe Syrquin, Simon Teitel, eds., Trade, Stability, Technology, and Equity in Latin America (New York: Academic Press, 1982), pp. 281-317.

21. Christopher Roper and Jorge Silva eds. Science and Technology in Latin America (London: Longman, 1983) contains a country by country directory of research institutes.

22. Ibelis Velasco, "Some Facts and Many Impressions on Science and Technology in Argentina," Part I, Interciencia, Vol. 8, No. 3 (May-June 1983), pp. 166-172; Part II, Vol. 8, No. 5 (Jul.-Aug. 1983), pp. 224-231.

23. Ibelis Velasco, "Some Facts and Many Impressions on Science and Technology in Mexico," Part I, Interciencia Vol. 6, No. 6 (Nov.-Dec. 1981), pp. 402-407; Part II, Vol. 7, No. 1. (Jan.-Feb. 1982), pp. 37-44.

24. Ibelis Velasco, "Some Facts and Many Impressions on Science and Technology in Chile," Interciencia, Vol. 9, No. 2 (Mar.-Apr. 1984), pp. 92-97.

25. Ibelis Velasco, "Some Facts and Many Impressions on Science and Technology in Costa Rica," Part I, Interciencia, Vol. 7, No. 3 (May-June 1982), pp. 166-169; Part II, Vol. 7, No. 4 (Jul.-Aug. 1982), pp. 236-239. Ms. Velasco is a Venezuelan science writer who received an IDRC grant in

1981-1984 to report on science and technology in several Latin American countries.

26. One of the most interesting accounts of applied research in Cuba is Charles Edquist, *Capitalism, Socialism, and Technology, A Comparative Study of Cuba and Jamaica* (London: Zed Books, 1985), which examines mechanization of sugar harvesting.

3

The Caribbean: Can Lilliput Make It?

Wallace C. Koehler, Jr., and Aaron Segal

The mobilizing of science and technology for development in the Caribbean is proving to be agonizingly slow. Although reliable information on research and development expenditures and research personnel is not available, the region and each of the member states remains overwhelmingly dependent on imported science and technology.[1] Efforts to foster indigenous capabilities are at very different states from country to country but their impacts are still limited. While rapid progress has been made in a number of countries science and technology remain marginal and precariously institutionalized.

There is no accepted and uniform definition of the Caribbean nor need there be. We define the region as consisting of the islands of the Caribbean Archipelago and the culturally related countries of Belize, Guyana, Suriname and French Guiana with the majority of their populations living on the Caribbean Sea. This provides in 1987 a region consisting of over 30 million people in 22 independent and non-independent countries speaking English, French, Spanish, and a variety of dialects and Creole languages. It is in this region that scientific and technological exchanges have existed for several decades and where a rudimentary regional S & T network is beginning to take shape. The five Central American republics and Panama operate essentially in another S & T framework although the Caribbean has much to learn from the impressive experience of Costa Rica.[2]

Our emphasis is on the development of indigenous capabilities for research, development, demonstration, adaptation and diffusion of science and technology (R,D,D,A, and D). The research to development cycle is further disaggregated in this definition to indicate the entire process and the stages at which Caribbean countries may participate. Thus most basic research and much applied research will continue to be imported but the region has a role to play in demonstration, adaptation, and diffusion. Indigenous capabilities are broadly defined to include research by multinational corporations or other non-regional actors provided that it is carried out in the Caribbean and is of relevance to regional needs.[3] Our interest is in the human resource capabilities of the Caribbean peoples.

Science and technology are used to make weapons, medicine, food, knowledge and many other items. Tradeoffs and contradictions between equity and efficiency goals, ecological and economic growth objectives, are persistent in the region.[4] Currently indigenous S & T is so limited that it makes a minimal contribution to any of these objectives, not even in Cuba which invests more in research than anyone else in the region. There is almost no military research in the Caribbean but there is also not enough of any other research to contribute significantly to economic growth. The evolution of indigenous capabilities can be measured in several ways including publications and citations in internationally circulating journals, patents and copyrights, R&D expenditures, cost-benefit analysis of research projects, quality of life indices, and air and sea pollution counts. Economic analysis suggests that one fourth to one half of economic growth in countries such as Brazil and the United States can be attributed to science and technology. The work of economist Nathan Rosenberg and others underlines the importance of shop-floor innovation, learning by doing, in the process of economic growth.[5] The scanty evidence indicates that the Caribbean has little formal or informal shop-floor R&D.

History of Caribbean Science and Technology

There is a long uneven history of science and technology in the Caribbean which remains to be documented. Science for several centuries was the prerogative of learned amateurs; botanists, naturalists, physicians and others. Technology was mostly imported and lightly adapted. Rarely was either institutionalized. A major Spanish scientific expedition was based in Cuba from 1795-1798 but neither the University of Havana or any other 19th century Caribbean university or academic academy found a secure place for science.[6]

The first significant Caribbean adaptations of science and technology occurred in the late 19th and early 20th century with the introductions of the steam engine, railway, and control of yellow fever and other mosquito-borne diseases. The striking decreases in mortality in Cuba, Puerto Rico and the West Indies after 1900 were the result of applied research, demonstration and diffusion.[7] These successes contributed to the establishment in the 1920's of modest agricultural, tropical medicine and public health research facilities.

In general the Caribbean colonial heritage in science and technology was meagre, largely oriented towards production of export crops, and failed to provide career opportunities for local scientists. Secondary and university education retained a humanities and law bias and predominant enrollment throughout the colonial period. Rigid race and class statified societies failed to diffuse popular knowledge of science and technology.

The drive towards indigenous science and technology capabilities has roots in Caribbean political nationalism. It is an expression of the desire to reduce political and economic dependency, to provide outlets for national creativity, and to generate economic growth which is subject to national direction. Caribbean Development Bank (CDB) President William Demas declared that "what Third World countries need is a vast increase in expenditure on Research and Development which would enable them to utilize their own domestic raw materials and ultimately to produce and export products based on their own resources or their own design styles. Even more important, technological innovation in Third World countries is required to develop efficient labour-intensive techniques of production."[8] The two themes of indigenous R & D for new exports and for appropriate technologies were linked to the desire to alter the terms of technology transfer.

The concern for national science and technology policies, planning and institutions began in Cuba in the 1960s and reached by the 1980s most of the region. The concept that science and technology required government force-feeding as well as regulation was promoted by several United Nations agencies, especially the Economic Commission for Latin America. This concept was fortified by the energy crisis of the 1970s and the felt need of governments to respond with coherent national energy policies. Conferences, seminars and workshops spread the message to politicians, civil servants and researchers. All independent Caribbean governments were asked to present national science and technology plans at the 1979 UN Conference on Science, Technology and Development. Most complied and for many it was their first attempt at a policy statement.

The new government awareness of possible roles for science and technology has not been accompanied by private sector or academic participation or much public support. Scientific communities within the Caribbean have vastly extended their formal and informal contacts over two decades but their principal ties are still outside the region. Lacking internal funding, adequate equipment, competitive salaries, technicians, and information services, most Caribbean national scientific communities are loosely structured and organized. At the regional level their ties are still embryonic. The pressure for mobilizing science and technology has come from the politicians rather than the scientists. It has come from the frustrations of energy imports, massive external debts, limited markets for traditional exports, and popular demands. It is often derived from a naive belief that science and technology once mobilized could provide responses to urgent short-term problems. At the 1983 first meeting of Caribbean ministers responsible for science and technology one politician remarked "I cannot go back to my Government and say that all we have produced is another report."[9]

The promise of a mobilized science and technology can only be realized if and when indigenous infrastructures come into being. This requires years of effort at improving and extending the teaching of science in the schools, popular science and technology education programs for adults, the establishment of critical masses of well-funded and supported researchers effectively networked within and outside the region, and agreement on research priorities. There are few shortcuts without an infrastructure and no shortcuts to its achievement although its size will vary. A quick review of national efforts to date conveys the state of existing infrastructures and research program.

National Efforts

Cuba is the only Caribbean state to have made research and development on sugar its primary concern. Three decades of expensive and intense efforts with Soviet help has produced mixed results. Although Cuba is the leading sugar producer in the world it does not export sugar technologies on a significant scale. The Cuban-Soviet designed mechanized sugar harvester built in Cuba since 1977 is inferior in productivity and more expensive to operate than

Western commercial harvesters.[10] Cuba has mechanized about fifty percent of its cane harvests through combinations of imported combines, its Cuban-Soviet made combine, and the extensive pre-burning of cane.

Cuba has the most impressive science and technology infrastructure in the Caribbean but it is not working well.[11] At the top is the Cuban Academy of Science which administers a dozen major institutes, science documentation centers, and S & T planning. Universities are relegated to training and some applied research while enterprises lack funds and authority to engage in shop-floor adaptation and innovation and learning by doing. The central research institutes are poorly linked to universities and to enterprises. Investment in R & D, especially sugar mechanization, does not appear to be fostering economic growth or reducing external dependency. The major Cuban equity gains in extending education, health, and other services have been achieved through management and investment, not R & D.

Puerto Rico has a science and technology infrastructure in search of a policy. Next to Cuba it has the largest number of researchers and research spending in the region. US federal government agencies support agriculture, forestry, fisheries, climatology, and other basic and applied research in Puerto Rico. The University of Puerto Rico and several other newer Puerto Rican universities carry out applied and basic research. The Puerto Rican government has modest applied research programs in a number of fields. While the private sector relies basically on unrestricted technology transfer from the United States, there is evidence that some informal shop-floor adaptation goes on in Puerto Rico.[12]

However, Puerto Rico has no national science and technology planning, policy or institutions. The Center for Energy and Environment Research of the University of Puerto Rico initiated a study of the viability of a science and technology center. As a consequence of the study the Governor appointed a commission to further consider the proposal. The commission recommended that a center be established as well as enhancing the research capabilities of the University of Puerto Rico, and more emphasis on the needs of small entrepreneurs. As of this writing, the report has not been officially acknowledged, in part because of a change in Governors. The plan would involve the use of fiscal incentives to motivate multinational firms located in Puerto Rico to substantially increase their local R & D efforts. It would be the first attempt in the Caribbean to establish institutionalized university-private sector links for research, drawing on US experience.

The Dominican Republic has fragmented and highly uneven research in agriculture, alternative energy systems, fisheries and other areas.[13] Government ministries, parastatal corporations, non-profit foundations, and the universities compete for far too few researchers, technicians, and funds. Efforts at coordination through science and technology offices in the Presidency and presidential science advisors have faltered. Each R & D unit seeks to jealously guard its turf. The National Energy Policy Commission was established in 1979 and has launched several research programs but with little coordination or coherence. If Cuba is overcentralized then the Dominican Republic has spread scarce resources too thinly and widely. It has particularly neglected investment in science education, science for adults, and science information systems. One result is that it is still basically dependent on overseas graduate study in the sciences and engineering in spite of huge increases in undergraduate student enrollment.

Haiti has for its 5 million population the weakest science infrastructure in the region. Three decades of brain-drain have resulted in more Haitian researchers abroad than within the country. A handful of foreign-funded projects in agriculture, alternative energy, and reforestation through fast-growing species go on but without an infrastructure. High turnover, low salaries, poor networking, no information systems, and other problems quickly frustrate researchers. National plans and policies are reduced to empty words in the absence of an infrastructure or serious efforts to create one. Since most Haitians receive less than 3 years of formal education, one must begin with elementary science concepts imparted by audio-visual, radio and other means in Creole rather than French which is not understood.

One of the few hopeful elements in the Haitian picture is the remarkable informal learning by doing of Haitian entrepreneurs in producing local components for assembly plants. Joseph Grunwald of the Brookings Institution recently conducted a study comparing backwards linkages in assembly plants in several countries. He found that Haiti's record was outstanding, taking advantage of low-cost labor, and tax and other incentives to replace imported with local components for baseballs and other products.[14]

The French Antilles and Guiana and the Netherlands Antilles still rely on metropolitan countries for most of their science, technology and institutions. This results in excellent marine biology, tropical forestry and other centers manned by European scientists. Applied research on local problems has had though to wait the organization recently of local universities and research institutes.

The Caribbean independent mainland states of Belize, Suriname and Guyana share low population densities, large tracts of undeveloped territory, and the possibilities of unexploited natural resource. Their research efforts and policies are at similar stages of seeking the funds, personnel and organization to carry out comprehensive natural resource surveys. Government ministries, parastatal organizations, and universities and technical colleges are unequal to the task and donors operate on a project by project basis. Guyana with its predominant public sector has gone furthest in national science and technology policy and planning but has little ability to implement. Belize and Suriname are mostly groping to improve extremely weak infrastructures.

The smaller Leeward and Windward Islands lack policy, planning, institutions, researchers, and research. Scattered projects are externally funded and implemented, often on alternative energy, with minimal local participation. The exceptions are the appropriate technology centers promoted by the Caribbean Council of Churches but their record of adaptation and diffusion of results is spotty. There has been little consideration of what constitutes appropriate science and technology infrastructure for these islands and too much emphasis on policy and institutions which are appropriate.

Perhaps the emphasis in the smaller islands of the Eastern Caribbean should be on science education and popular science for adults. Long-distance teaching by radio and satellite, computer and audiovisual technologies can all be used to raise indigenous capabilities without costly formal instruction. Research should be undertaken at the request of and with the full participation of locals even if this means a slower research timetable.

There is an enormous contrast between the R & D capabilities of Trinidad and Tobago and those of the rest of the Eastern Caribbean. Housing a University of the West Indies campus, the Caribbean Industrial Research Center serving the private sector, a branch of the Caribbean Agricultural Research Development Institute, and various government ministry efforts, Trinidad has a working if inadequate infrastructure. The government decision to invest oil revenues in joint venture industrial export projects in petrochemicals has also improved local information and documentation capabilities. Trinidad has and should continue to provide advice on technology and technology transfer to the Eastern Caribbean.

Like Puerto Rico, Trinidad has an infrastructure in search of a policy. This is reflected in the discussions over a strategy of joint ventures and technology transfers, industrial import substitution, and the proposed National Institute of Higher Education, Research, Science and Technology. Small-scale scattered applied research efforts in a number of areas including agriculture and marine biology have limited impacts. Attention is needed to science education and information to improve and extend the infrastructure.

Barbados has relied on informal and formal networks to achieve coherent if modest performance. It benefits from the location in the country of the Caribbean Development Bank, the headquarters of the Caribbean Meteorological Institute and other regional organizations with technical capabilities, including the local campus of the University of West Indies. It has achieved some success with commercial dissemination of work on biogas digesters, solar heaters, and agro-industry. It has also recently surveyed its research, researchers, and spending and has baseline data generally absent elsewhere. The role played by universal literacy, public awareness of S & T, and informal public-private sector linkages has given Barbados an edge. The question may be whether to continue with effective gradual efforts or to attempt more rigorous and concentrated priorities and performance.

Jamaica has had a topsy-turvy experience with science and technology in recent years including a stark exodus of professionals and technicians in the 1970s, and a drastic switch from emphasis on controlling the transfer of technology to encouraging uncontrolled transfers. There have also been numerous changes in personnel in institutions responsible for science and technology. What has continued is a basic and applied research capability at the Jamaica campus of UWI; especially at the Medical School and the Caribbean Food and Nutrition Research Institute; a tradition of government research in agriculture as well as private efforts, and some scattered energy, fisheries, and other R & D. A key problem is too many small, uncoordinated research efforts underfunded and understaffed.

Jamaica has severe infrastructure and policy problems. It must provide competitive salaries and working environments which probably means regrouping researchers in groups of minimum efficient size. Cooperation between public and private sector is essential if research is to be adapted and diffused. Consideration of fiscal incentives for R & D is relevant in an economy crippled for lack of foreign exchange.

The College of Science and Technology has a useful role to play in working with the private sector to foster shop-floor innovation and training. A national policy and plan may be appropriate for Jamaica if the process is open and participatory including the increasingly organized scientific community.[15]

These thumbnail sketches of national efforts are partial, subject to change, and arbitrary. They do indicate the enormous range of science and technology experiences and approaches within the region, and the basic obstacles to regional cooperation. Such cooperation at present consists of the Caribbean Community (CARICOM) nations whose relations focus on politics and trade but also includes, UWI, CMI, CDB, the Caribbean Examination Council, and a number of non-governmental professional associations. At the regional level the Association of Caribbean Universities and Research Centers (UNICA) founded in 1967 has continued a low-profile program of conferences, workshops and exchanges of information and has discussed possible joint research projects. Its membership includes universities throughout the Caribbean, as well as Colombia, Venezuela, Mexico, and the US, but Cuba has not joined.

The Commonwealth Caribbean has attempted several regional science and technology projects and proposed others. Using US funding, the Caribbean Development Bank and the CARICOM Secretariat have spent $7 million over five years on small-island alternative energy research. The CDB also operates a Technological Consultancy Service for the Eastern Caribbean. The Organization of American States has had several small-scale subregional projects. The CARICOM Secretariat lacks the authority and the technical competence to coordinate these efforts. CARICOM at the political level appears to be too beset with major problems and too unwieldy to place attention on S & T initiatives.

Instead the focus since 1979 has been at the Caribbean-wide level with the initiative coming from ECLA and UNESCO and a few individuals such as Dr. Dennis Irvine, former Vice-Chancellor of the University of Guyana. These efforts produced in 1981 the intergovernmental Caribbean Council of Science and Technology (CCST). Its membership includes most of the CARICOM states plus Cuba, the Dominican Republic, Haiti, and even the Netherlands Antilles as a possible non-independent Associate Member. It is the widest Caribbean governmental grouping for science ever except for the World War II and postwar Caribbean Commission that was confined to the colonial powers. However, several CCST members have not paid their dues, lack of internal and external funding has continued reliance on ECLA for Secretariat services, and member participation and interest is markedly uneven. There is agreement on a specific "coordinating, advisory, and implementation role" for the CCST.[16] The initial work program calls for a regional science journal, assessment of national S & T capabilities, and other information and exchange activities. Like UNICA, the CCST with such a diverse membership, has settled for activities likely to afford benefits to all if at a lowest common denominator.

The state of regional and sub-regional activity is growing but still incipient. The extraordinary range of bilateral and multilateral donors certainly results in duplication, fragmentation, and too many donors chasing too few qualified researchers. Regional and subregional cooperation is easiest at the level of exchanges of information and yet to be realized at the level of joint research or support of research centers except in the Commonwealth Caribbean. The dilemma is that without much more extensive regional cooperation many Caribbean countries will be shut out of science and technology.

The present state and prospects for S & T in the region need to also be examined by sectors. Table 3.1 provides available information on current national and research spending; a more reliable guide than policy statements. There is striking convergence and an apparent basis for further regional cooperation. Our discussion attempts to highlight the issues in each key research sector.

Alternative Energy Research

The Caribbean is 90 percent dependent on imported oil at present to fuel its energy needs (Trinidad and Tobago is the only oil and gas exporter, Barbados and St. Vincent produce some oil and gas). Yet the Caribbean and other sub-tropical islands have energy advantages not necessarily shared by other developing countries. The energy opportunities associated with coastal activities are of particular interest.[17]

It is generally recognized that the Caribbean possesses a wide array of energy resources which may be exploited to provide from a small to a large proportion of indigenous

TABLE 3.1
Current Caribbean Research and Development Spending Priorities

Country	Priorities
Barbados	Alternative energy, sugar and byproducts, crops, food processing
Belize	Agriculture, fisheries
Cuba	Sugar and byproducts, industrial technology, pharmaceuticals, electronics, construction
Dominican Republic	Alternative energy, hydro, sugar and byproducts, fisheries
French Antilles &	Space, agro-industry, marine biology
French Guyana	Alternative energy, agriculture, forestry, minerals
Haiti	Agriculture, alternative energy, forestry, assembly plants backwards linkages
Jamaica	Alternative energy, agro-industry, crops, construction
Netherlands Antilles	Alternative energy, marine biology
Puerto Rico	Alternative energy, agriculture and agro-industry
Suriname	Alternative energy, hydro, agriculture, fisheries
Trinidad and Tobago	Industrial technology, agro-industry, fisheries

(Continued)

TABLE 3.1 (Continuation)

Regional	
CDB/CARICOM	Funding alternative energy, technological consulting
CADEC	Appropriate technology centers
CARDI	Agriculture and forestry
CCST	Coordinating science and technology policy and publish science magazine
UNICA	Hold agriculture and energy university research workshops
UWI	Basic and applied research at Barbados, Jamaica, and Trinidad campuses
CMC	Collect meterological data for Commonwealth Caribbean

Sources: Report of the First Meeting of Ministers Responsible for Science and Technology, Chairman Dr. R. A. Irvine, Kingston, 6-7 April, 1983 and *Interciencia*, 1979-1984, Inter-News.

energy needs. (Table 3.2) Renewable energy presents the greatest opportunities. Recent oil and gas explorations in Cuba, Jamaica, Guyana, and Suriname have yet to produce significant finds, Puerto Rico and the Dominican Republic may have offshore reserves; however prospects elsewhere are slim. By contrast there is extensive solar insolation, the winds tend to be strong and predictable, good ocean thermal potentials exist, several countries have geothermal and/or hydro resources, and the biomass resource base is large and varied. In spite of lower world oil prices the extreme dependence of the Caribbean on imported energy remains a severe economic burden, and research on alternative energy an important opportunity.

There is disagreement over the appropriateness of research, development, demonstration, application and diffusion focusing on renewable energy. Some analysts favor a wide range of research programs.[18] Others come up with priorities and propose development by external sources using economically and technically proven technologies and donor-imposed regional, subregional and national energy policies.[19]

The track record of energy research in the region is mixed to date. There is a lengthly list of donors, projects, and sectors which includes foreign governments, international organizations, private foundations, and others. Some governments have responded by organizing their own national policy offices as in the Dominican Republic, Puerto Rico and elsewhere.

In spite of this activity and interest, there has been relatively little actual energy research in the region. The Center for Energy and Environment Research in Puerto Rico has been the single most active research center, working on energy from sugar cane, solar air conditioning, industrial hot water, ocean thermal energy and other technologies. Because of changing US government priorities the Center has had to curtail much of its work. The CDB has funded a variety of research, including a passive solar water heater program in Barbados. It too has run into funding constraints on future energy research. The Regional Energy Action Plan proposed by the Organization of Eastern Caribbean States is problematical due to lack of external funding. The first round of energy research risks being lost or dissipated if the donors lose interest or change priorities.

The goal of reducing energy dependency has been widely accepted but not translated into action. Recognition that energy research requires long-term commitments to

TABLE 3.2
Development and Potential of Energy Resources in the Caribbean

Island or Country	Oil and Gas	Coal	Hydro Power	Geothermal Energy	Biomass Energy	Solar Energy	Others (Wind,etc,)
Antigua	1A	1A	1A	2A	2A	5A	5A
Bahamas	2A	1A	1A	2A	2A	5A	5A
Barbados	3B	1A	1A	2A	4B	4A	5A
Colombia	4D	1C	5D	2A	5B	4A	5A
Cuba	3C	2A	3B	2A	5B	5A	5A
Dominica	1A	1A	4C	2A	2A	4A	5A
Dominican Republic	2A	2A	3B	2A	5A	4A	5A
Grenada	2A	1A	2A	2A	2A	4A	5A
Guyana	2A	1A	2A	2A	2A	4A	5A
Haiti	2A	1A	3B	2A	4A	5A	5A
Jamaica	2A	2B	3B	2A	5B	5A	5A
Martinique	1A	1A	1A	2A	4B	4A	5A
Mexico	5D	5C	5C	4C	5B	5A	5A
Monserrat	1A	1A	1A	2A	2A	4A	5A
Puerto Rico	1A	1A	3B	2A	5B	5C	5A
St. Kitts-Nevis	NA	NA	NA	NA	NA	NA	5A
St. Lucia	1A	1A	1A	3A	2A	4A	5A
St. Vincent	1A	1A	3C	2A	2A	4A	5A
Trinidad/Tobago	5A	1A	1A	2A	3B	4A	5A
Venezuela	5D	2B	5C	2A	4B	4A	5A

POTENTIAL

1. Poor
2. Not determined but possible
3. Limited
4. Medium
5. Important

DEVELOPMENT

A. Without development
B. Limited development
C. Medium development
D. Good development
NA = Not Available

DATA FORM

Esquema de la energia y el ambiente en la zona del Caribe, 7 de agosto de 1979, Organizacion de las Naciones Unidas.

infrastructure in order to train, retain, and retrain qualified researchers has often been missing. Discussions of international, regional and national planning, policy, and cooperation skip the specifics needed to sustain energy research. Project by project episodic funding makes it difficult to develop those very indigenous research capabilities that are needed.

Agriculture and Forestry

Export crops such as sugar and sea-island cotton have provided the historically most effective examples of Caribbean public and private sector research linkages. Discouraging markets and prices for traditional exports present new challenges to a post-colonial research structure.

There are advocates of new research programs on non-traditional export crops such as fruit trees for whose products new markets may exist.[20] The emphasis is placed on commercialization and marketing. Others maintain that research should focus on low-cost, labor-intensive technologies at the disposition of small farmers with little credit or formal education. Then there are those who argue for agro-industry research to adapt known dairy, poultry, sheep and pig, animal fodder and other conditions to Caribbean commercial agriculture and food processing. The emphasis here is on agricultural extension, mechanization, and technology transfer with the goal of reducing present extremely high food imports.

The debate over research approaches and goals divides governments, ministries of agriculture, researchers, university faculties of agriculture, external donors and others. It even occurs in Cuba where the small remaining private sector is denied research but still outyields the state farms.[21] It is a debate with a different balance in each country due to the different prevailing systems of land tenure, extent of rural migration, and other factors. For instance Puerto Rico has opted for agro-industry research in a society where few smallholders remain; Haiti is overwhelmingly rural and small farmer and concentrates on labor-intensive research. The debate is further complicated by the possible use of sugar for fuel and its economics.

The problem is that at the national level the resources are lacking to effectively pursue several agricultural research strategies at the same time. A World Bank study of developing country agricultural research has indicated the diseconomies of scale from too few and isolated researchers. Work on new crops and traditional crops such as sugar and bananas must be carried out at the subregional or regional level for the smaller countries. Given a regional division of research labor it might be possible to follow several research lines simultaneously but this is a long-term goal.

Unfortunately it appears that research decisions need to be forced between helping smallholders or agro-industry. A similar although less painful decision lies between research on commercial forestry and fast-growing species for reforestation in peasant societies. Haiti and the Eastern Caribbean must choose agricultural and forestry research for peasants while the rest of the Caribbean de facto opts for agro-industry. Ironically agro-industry research is less expensive because it involves adapting proven large-scale technologies through scaling-down. Since the Green Revolution was for cereals and rice there is no on-the-shelf technological package for tropical small farmers and much costly and time-consuming basic research is needed.

Appropriate Technology

The concept of labor-intensive, small-scale technologies have received an enthusiastic reception in much of the Caribbean. Church groups, non-profit foreign donors, and other organizations have sponsored centers, fairs, meetings and demonstrations. Results are uneven and mixed but an important increment to adult technology awareness and skills has occurred, especially in the small islands. The appropriate technology groups have also developed formal and informal networks and information-sharing; an important lesson for the scientific community. While its total economic contribution may be limited, appropriate technology efforts in the region are a welcome sign of self-reliance. Where local interest merits appropriate technology efforts may be extended to crafts, construction technologies, materials recycling, and small industries.

Environmental Sciences

The Caribbean consists of densely populated highly fragile human and organic ecosystems subject to periodic hurricanes, earthquakes and manmade disasters such as oil spills. The environmental sciences are recent arrivals in the region although there is a distinguished record of academic research in marine biology in Puerto Rico, Trinidad, Jamaica, Barbados, Curacao, and elsewhere. Recognition of environmental concepts has been stressed by UNEP, UNESCO with its Man and the Biosphere research program, and by the non-governmental Caribbean Conservation Association. Ecological problems have also received some attention from the Caribbean Tourism Center in Barbados established by the Caribbean Hotel and Tourism Association.

The growth versus pollution debate of the 1960s and 1970s has a different context in the Caribbean. Pollution in a closed island ecosystem threatens survival in a way that it does not in Calcutta or Mexico City. There has been growing demand for applied research on short-term problems of harbor pollution, oil-spills, coastal zone management and beach and sand erosion and coastal and fish farming. There are political demands for research to improve fishing practices and yields, reduce imports and generate employment.

Unfortunately increased interest in ecological research has not been matched by a strengthening and revision of environmental science infrastructures. Technicians are desperately scarce making fisheries and marine extension programs unrealistic. Research centers lack critical masses of researchers, and adequate information services with a consequent loss of staff. Important work has been done in Caribbean archaeology, marine biology and other fields but often through collaboration between local and better-equipped foreign researchers. The small islands have become particularly dependent on donors for assistance with their multiple ecological problems. The possibility of regional cooperation immediately runs into the short-term needs of many countries versus the long-term commitment of building infrastructure.

Climatology and seismology are the two disciplines in which Caribbean applied research and international basic research interests have been bridged. The Caribbean Meteorological Institute collects weather data for the Eastern Caribbean and uses satellite data for forecasting and hurricane and storm warnings. Its cooperation and that of other Caribbean national weather services with US agencies has markedly improved regional forecasting capabilities while adding to global data. Similarly international oceanographic and seismic work on the Caribbean has added to basic research knowledge of tectonic plates and planetary climatic history.[23] The lesson is that the Caribbean can participate in first-rate basic research by providing facilities and staff and matching applied to basic research interests. The principal advantage comes from the on-the-job training of Caribbean researchers.

Industrial Research

There is very little formal industrial R & D in the Caribbean and an unknown but presumably limited amount of informal shop-floor adaptation. Cuba is the sole exception with its need to adapt Soviet and East European capital goods and its efforts at industrial import-substitution including designing its own micro-computers. Elsewhere technology transfer is largely unregulated except for foreign exchange constraints. Industrial technology institutes in the Dominican Republic, Trinidad, and Jamaica provide information services, consulting, and some trouble-shooting for the private and public sectors.

The debate over industrial R & D in the Caribbean has several dimensions.[24] One element concerns the range of choices and terms of technology transfer and calls for regional or other advisory mechanisms. The problem is that the same scarce pool of manpower is available for research or for assessing technology to be transferred. Another element concerns the need for regional design and feasibility capabilities for new export industries such as petrochemicals.[25] Again the lack of available manpower suggests that the costs of establishing such capabilities would come at the expense of other R & D. There is also the element of fostering local adaptation in industrial import-substitution rather than simply scaling-down technologies. There is room here for experiment with fiscal incentives to encourage energy conservation and other forms of adaptation in plants producing for local or regional markets. Finally there is the need to promote backwards linkages in assembly plants in order to increase employment, taxes and use of local materials. This opportunity merits regional study and use of fiscal incentives.

Information and Social Sciences

There has been more than 50 years of solid scholarship in the social sciences in the Caribbean, much of it by local scholars. Topics such as race and class, kinship and gender, Africanisms in the New World, the plantation economy, emigration and others have been competently studied over several decades. The research findings have been widely diffused and constitute part of the basic world views of many Caribbean people. There are a number of social science research centers in the region, notably the Institute of Social and Economic Research of UWI, and a steady stream of publications.

While research must continue on the topics first delineated before World War II there are signs of new emphases. Management of enterprises-public, private, non-profit, cooperative, etc. urgently requires understanding in these societies. Urban planning, land use, coastal resource management, are intellectual imperatives for research in the face of rapid change. Researchers need to come to grips with tourism as a multidisciplinary phenomena requiring highly sophisticated research rather then superficial analysis. Longitudinal and cross-cultural research which treats the entire region as an entity has yet to be realized. As science and engineering research in the region increases the social sciences which have played a leading role need to expand their interests and empirical data bases.

Health Sciences

The Caribbean strength lies in applied research such as drug trials, demonstration and diffusion. Basic research on tropical medicine continues in Cuba, Jamaica and Puerto Rico but major advances are likely to be made elsewhere. Instead the challenge is to devise and implement para-medical health delivery systems in those countries where universal hospital and physician based medicine is not feasible. These are essentially public health and management challenges and there are important gains from regional sharing of information and comparative research.

Natural Resources R & D

Several Caribbean countries such as Guyana, Belize, Suriname and French Guiana have extensive unexplored areas of potentially economically valuable natural resources. Other countries such as Cuba and Jamaica have potential oil and gas deposits. Natural resources research also includes uses of local materials such as kaolin, minerals processing and marketing studies. This research tends to be expensive and highly risky. Investment in regional capabilities is not justified except perhaps to participate in joint ventures. Where this has occurred as with the state petroleum corporations of Jamaica and Cuba it is not clear that an increase in useful indigenous capabilities has taken place.

Service Sector

The most developed Caribbean economies have growing service sectors, even if their industrial bases are limited. There has been no systematic research on service sector productivity in the region although this may be an important factor in future economic growth. Issues of office automation, industrial relations in the service sector, banking productivity, retail and wholesale trade organization and others merit attention. The tourism sector has yet to be studied from a productivity perspective. Wages policies in the service sector need to be examined also in relation to motivating output. As the balance shifts from agriculture to assembly plants to tourism and services so do the relationships of individuals to technologies. There is no study of the comparative use of computers in the Caribbean.

Human Resources

The Caribbean for two decades has barely been able to replace its existing numbers of researchers and in several countries such as Haiti there are fewer researchers now than there were in 1960. Investment in science education and science teaching at all levels is the highest priority due to the long lead-times needed to train researchers. Augmentation of science education with fairs, clubs, prizes, science museums, audiovisual materials, etc. is vital and lends itself to regional cooperation. Science education for adults is also important on the job, through clubs, unions, and other organizations. The goal should be augmented job-related knowledge and skills rather than a vague awareness of the importance of science. Audiovisual and computer on the job training should be attempted.

Numerous studies have shown that researchers emigrate due to frustration with local working conditions and salaries as well as foreign opportunities.[26] The Caribbean has the advantage of geographic proximity to major research centers and possible on-line communications. Keeping good researchers in the region requires providing them with frequent keep-up access to major centers, on-line data bases and overseas communications, and centers with "critical masses" sufficient to permit stimulating exchanges. Handfuls of isolated researchers scattered around the region are not productive. Adequate information systems and telecommunications are a sine qua non of effective Caribbean R & D; not luxuries. The alternative is to continue to see some of the best people emigrate.

Research Priorities

Several lists of possible Caribbean research priorities have been put together.[27] Ours is derived from the long-term goal of building indigenous research capabilities. It argues for highest priority to *in situ* research on problems unique to the region where transferable technologies will not work or must be adapted. Renewable energy systems and agriculture and appropriate technology fit this criterion. So does research on Caribbean ecosystems. Investments in information science, improved telecommunications and science education are needed to make any R & D program possible, including our suggestions. The priorities we propose require infrastructure buildups and cannot promise economic results before the 1990s. We do not believe that there are short-cuts in the Caribbean. Science and technology in the region must be nourished before it can deliver. Short-term crash projects lead nowhere since local capabilities are not altered. Continued reliance on technology transfer cannot deal with sectors where *in situ* research has no substitute. It is possible to argue for other priorities but a minimum 10 year time-frame is essential. Otherwise researchers and centers will be asked to deliver what they cannot and disillusionment will be general.

Approaches

Donors to Caribbean science and technology have their individual agenda and constituencies. The World Bank has sought with some success to coordinate major government donors. An indication of broad funding levels for several years in advance would help. It is undesirable though for donors to dictate priorities or to coerce clients into regional or subregional cooperation. The donors can insist though that clients match stipulated priorities with their own resources. Currently a majority of Caribbean R & D is directly externally funded in every country except Cuba.

Indigenous capabilities need to be increasingly funded from indigenous resources.

Most R & D in the Caribbean will continue to be carried out at the national level, whatever the sources of funding. Funding needs to be restructured to facilitate user-researcher linkages. Fiscal incentives can be tried to induce the tourist sector to fund solar energy; agro-industry to support university work, etc. The self-imposed segregation of researchers and possible users must be forcefully broken-down or no diffusion will occur. Where national councils of science and technology exist there should be broad participation of trade unionists, farmers groups, teachers, etc. The smallness of these societies should be an asset for research diffusion and not a liability. Public sector corporations like the electric utilities should have set aside R & D funds to be used for contracting with universities and the private sector. Linkages should explicity aim at strenghtening local and regional engineering and design capabilities. Non-profit organizations also have an important role to play in R & D support. The donors can create supply but demand for research is a function of linkages established nationally.

The scope for regional and subregional cooperation is extensive; the prospects so-so. Even Cuba, Puerto Rico, Trinidad, and Jamaica will within a decade exhaust the R & D they can effectively perform at an island and national level. The smaller countries participation rules out projects of most interest to the most advanced. Bilateral cooperation as between Puerto Rico and the Dominican Republic or Cuba and Jamaica as in the 1970s may be more promising but can also be unbalanced.

Donor and internal support is needed to maintain the momentum in support of regional cooperation. There needs to be a step ahead from conferences and surveys to carefully designed shared research. It is true that baseline data is unavailable on most countries and that science and technology policies are often lacking or non-existent. What does it mean though to ask a government which has no research or researchers to produce a policy? It is better to step on the infrastructure and R & D acclerator. The convergence of current research agendas and spending patterns indicates the possible gains from launching regional projects.

Conclusion

How to get from nowhere to somewhere? The Caribbean at

present does not have sufficient science and technology capabilities to effect its own future. Compare this to India which was able to demonstrate, adapt, and diffuse the Green Revolution to change from a net food importer to being food self-sufficient. Compare this to Singapore which has developed the ability to increasingly design and produce its own industrial exports. It is possible for the Caribbean within a decade to have the indigenous capability to alter its future in energy, agriculture, and ecology. This does not mean that these capabilities will be used or used wisely. Nor does it mean that all Caribbean societies will share in those capabilities, even if some are regional. Nor does it mean that dependency on imports will be necessarily reduced although the import mix could be changed. Surely it is better to import computers rather than apples and dried fish?

The alternative is also visible. It is a perpetuation of the status quo. Most energy is imported depending on the vagaries of world markets, prices, and politics. More and more food is imported and more and more rural people leave for Kingston, Portau-Prince, Miami or New York. Ecological pressures increase, more beaches erode, forests denude, and finite natural resources dwindle. The alternative is not apocalyptic but it is not pleasant. Science and technology do not have the answers to the outstanding problems of the Caribbean but they tell us how to look.

NOTES

1. There has been no regional survey of R & D and research manpower using a uniform methodolgy. Francisco R. Sagasti, "Ciencia y tecnologia en America Latina", Comercio Exterior, Mexico, December 1984, pp. 1163-1180 relies on several sources and estimates for different years. Figures for expenditures do not often indicate sources of funding. Baseline data on R & D expenditures by sector, numbers of full-time researchers, and other indicators remain to be collected for the region and for most of the individual countries.

2. Ibelis Velasco, "Some Facts and Many Impressions of Science and Technology in Costa Rica", Interciencia, Vol. 7, No. 4, July-Aug. 1982, pp. 236-40. The establishment of researcher-user ties is particulary relevant to the Caribbean.

3. The European Space Agency launch center in French

Guiana is the largest research project currently in the region although the relevance of its work is mostly elsewhere. See Frank Schwarzbeck, "Recycling a Forgotten Colony", pp. 22-25, and Gerhard Drekonja-Kornat, "On the Edge of Civiliazation", pp. 26-28, in Caribbean Review, Vol. XIII, No. 2, Spring 1984.

4. Fuat Andic, "Efficiency vs. Equity, Economic Policy Options in the Caribbean", Caribbean Review, Vol. XIII, No. 1, Winter 1984, pp. 16-20; Carl Stone, Power in the Caribbean, (Philadelphia: Institute for the Study of Human Issues, 1985) discusses Caribbean political systems in relation to growth and equity objectives.

5. Nathan Rosenberg, Inside the Black Box: Technology and Economics, (New York: Cambridge, 1982). Argentine economist Jorge Katz directed a series of studies of shop-floor technology in several industries and countries in Latin America. See his essay, "Technology and Economic Development: An Overview of Research Findings" in Moshe Syrquin, Simon Teitel, eds., Trade, Stability, Technology and Equity in Latin America (New York: Academic Press, 1982), pp. 281-317.

6. A comprehensive history of science and technology in the Caribbean from colonial times to the present has yet to be written. However important studies provide a partial picture. See Iris Engstrand, Spanish Scientists in the New World (Seattle: University of Washington, 1981), pp. 159-172; Manuel Moreno Fraginals, The Sugarmill (New York: Monthly Review, 1976) for a detailed account of the introduction of innovations to the Cuban sugar industry from 1760-1860; and especially Richard B. Sheridan, Doctors and Slaves, A Medical and Demographic History of Slavery in the British West Indies 1680-1834 (New York: Cambridge, 1985). Sheridan notes that "unlike the United Kingdom and the United States, the British West Indies failed to develop a medical culture of professional schools, hospitals, societies, and journals." p. 70.

7. J. Diaz-Briquets, The Health Revolution in Cuba (Austin: University of Texas, 1983), pp. 35-53. Similar rapid reductions in mortality followed the introduction of public health measures in Puerto Rico, and the British West Indies. The development of Caribbean sugar cane-breeding capabilities based on imported varieties in another early 20th century example of quick technology transfer.

8. William Demas, "How to be Independent," Caribbean Review, Vol. VI, No. 4, pp. 12-13.

9. Report of the First Meeting of Ministers Responsible for Science and Technology, Kingston, Jamaica, 6-7 April, 1983.

10. Charles Edquist, <u>Capitalism, Socialism, and Technology, A Comparative Study of Cuba and Jamaica,</u> (London: Zed Books, 1985). Comprehensive studies of Cuban science and technology are not available. Limited or partial accounts include Tirso W. Saenz y Emilio Garcia Capote, "Ernesto Che Guevara y el Progreso CientificoTecnico en Cuba," <u>Interciencia</u>, Vol. 8, No. 1, Jan.-Feb. 1983, pp. 10-18; Marcel Roche, "Cuba: El Centro de Investigaciones Biologicas," <u>Interciencia</u>, Nov.-Dec. 1985, Vol. 10, No. 6, pp. 299-300; and Angela Tomeu Miranda, Cristobal Ellipe Gonzalez y Soledad Diaz Otero, "Ciencia, tecnologia y empleo en el medio rural cubano," in Viviane B. Marquez ed., <u>ciencia, tecnologia y empleo en el desarrollo rural de america latina</u> (Mexico: El Colegio de Mexico, 1938), pp. 261-286.

11. The problems of low-productivity in the Cuban economy are frequently cited by Fidel Castro and other leaders. They are discussed in Carmelo Mesa-Lago, <u>The Economy of Socialist Cuba</u> (Albuquerque: University of New Mexico, 1981), pp. 179-182.

12. The Center for Energy and Environment Research of the University of Puerto Rico conducted in 1984 a survey of industrial R & D in Puerto Rico. The results indicate very little formal R & D, mostly by Puerto Rico based companies, with somewhat more adaptation and dissemination.

13. Sources for impressions of the Dominican Republic and other Caribbean countries except Cuba include personal visits by the authors, interviews, the Internews section of <u>Interciencia</u>, the 1981 and 1982 reports to UNESCO on "Science and Technology for Development in the Caribbean: Current Status and Possibilities for Regional Cooperation" by Dr. D. H. Irvine. Latin American Newsletters, <u>Science and Technology in Latin America</u> (London: Longman, 1983) has brief sections on Belize, Cuba, the Dominican Republic, French Guiana, Haiti, Jamaica, Puerto Rico, and Suriname.

14. Joseph Grunwald and Kenneth Flamm, <u>The Global Factory</u> (Washington: Brookings, 1985), pp. 187-204 on Haiti.

15. The Jamaican Society of Scientists and Technologists and the Trinidad and Tobago Scientific Association are the only Caribbean members of the Inter-American Association for the Advancement of Science. There is also evidence of shop-floor R & D in Jamaica, especially at the ALCAN company. Personal communication.

16. Report of the First Meeting of Ministers Responsible for Science and Technology, Kingston, Jamaica, 6-7 April, 1983.

17. Juan A. Bonnet, Jr. and Wallace C. Koehler, Jr.,

"Status of Energy Programs in Caribbean Islands," Proceedings, Energex Conference, Regina, Sasketchewan, Canada, May 1984.

18. Ibid, and Latin American Plan for Action for the UN Conference on New and Renewable Sources of Energy, A/Conf.100/7, March 28, 1981 and A.Conf. 1008, April 1981, Draft Action Plan for the Caribbean Environment Programme, UNEP/CEPAL-WE 48/3, September 16, 1980.

19. Energy Resources within Caribbean Development and Cooperation Committee Member Countries, E/CEPAL/CDCC/65, May 28, 1980; and J. Vardi, Coordination of Energy Policy in the Caribbean, UNDP, Revised Report, June 1, 1982.

20. Board on Science and Technology for International Development, Needs in Science and Technology for Development in Caribbean Island Nations, Report of a Workshop, August 30-31, 1982. (Washington: National Academy Press), pp. 32-41.

21. Carmelo Mesa-Lago, The Economy of Socialist Cuba (Albuquerque: University of New Mexico, 1981), pp. 138-139.

22. World Bank, Agricultural Research, Sector Policy Paper, Washington, DC, June 1981.

23. Jorge Enrique Corredor "Identificacion y Analisis de Ecosistemas del Caribe," Interciencia, Vol. 9, No. 3, June 1984, pp. 145-152.

24. The literature on technology policies for industry includes a special issue of Social and Economic Studies, Special Number, Vol. 28, No. 1, March 1979 published by the University of the West Indies, Jamaica. See also Norman P. Girvan, Technology Policies for Small Developing Economies, A Case Study of the Caribbean (Jamaica: University of the West Indies, 1983) which emphasizes regulation of technology transfer; and Andrew Axline, Caribbean Integration (New York: Nichols, 1979), for a discussion of problems of regional cooperation in industrial policies.

25. Steve de Castro "A Technology Policy for Petrochemicals in CARICOM," in Social and Economic Studies, Vol. 28, No. 1, March 1979, pp. 282-336.

26. Ransford W. Palmer, Problems of Development in Small Beautiful Countries (Lanham; Maryland: North-South Press, 1984), pp. 30-38 on Jamaican emigration. The extensive literature on Caribbean emigration since World War II includes only a few studies which focus on the "brain-drain" or highly skilled emigrants. See Rosemary Brana-Shute, ed., A Bibliography of Caribbean Migration (Gainesville: University of Florida, 1983).

27. One of the most interesting proposals to restructure research is by Lawrence A. Wilson, Dean of Agriculture,

University of the West Indies-Trinidad, in "Toward the Future: An Alternative Framework for Agricultural Research, Training, and Development in the Caribbean," (St. Augustine: University of the West Indies, February 1984). Dennis Irvine, formerly Vice-Chancellor and Rector of the University of Guyana, presents his research priorities in his 1981 and 1982 reports to UNESCO on regional cooperation in science.

4

The Middle East: What Money Can't Buy

Aaron Segal

Science and technology in the contemporary Middle East from Morocco to Iran constitutes an extraordinary paradox. A handful of oil-exporting countries (Saudi Arabia, Kuwait, Libya, Algeria, the United Arab Emirates) have been annually spending billions of dollars to import state of the art science and technology for military and civilian purposes and foreign scientists and technicians to transfer these technologies. Meanwhile the majority of Middle Eastern states (Egypt, Turkey, Morocco, Sudan, Tunisia, Yemen, Jordan, Syria, Lebanon, et al) are struggling with varying degrees of success or failure to establish national science and technology capabilities. The many efforts at regional cooperation in science and technology between oil-exporters and the others have produced modest results to date. Israel alone in the region has achieved a significant indigenous research and development (R & D) capability, assisted by external support but also making its own ongoing contribution to original world science and technology. (S & T)[1]

This chapter focuses on the development of indigenous R & D capabilities rather than the terms or nature of technology transfer. It argues that without such capabilities, whether utilized for military or civilian purposes, no effective transfer can be said to have occurred. It adopts the conceptual framework of the economist Nathan Rosenberg that development consists of technological "learning by doing"; the ability to adapt, improvise, innovate, modify, and eventually to re-export imported technologies.[2] This capability is a result of both formal learning in universities and research centers and informal shop-floor learning through engineering maintenance, re-tooling, reprocessing, redesign, reverse engineering, and other production techniques. Learning by doing can occur in a rudimentary machine shop or a government munitions factory; a university department of engineering or a rural borehole rig. In its absence imported technologies breakdown and are abandoned and productivity is a hostage to the availability of foreign parts and experts.

Relevance of the Past

Any discussion of current Middle East science and technology must briefly examine the relevance of the historic Islamic scientific past. Dr. Ibrahim Madkour, President of the Academy of Arabic Language in Cairo, expresses a widely held view that at its apogee Arab science and technology "helped pave the way for the European Renaissance" before experiencing its own four centuries of "Dark Ages."[3] The proponents of this view see in the recognition of the historic role of Islamic science and technology the key to future greatness. However they are unable to establish the connections between real past achievements and present and future science and technology policies.

The rise, diffusion, height, and subsuquent decline of Islamic science and technology are the subject of an extensive literature.[4] The strengths of Islamic science in agronomy, botany, mathematics, optics, geography, medicine and other fields were real and a major contribution to the growth of world science. These strengths it is argued by economic historian Andrew M. Watson were primarily in description, collection, cataloguing, translation, and diffusion not in experiment or research. He maintains that "the advance and decline of Islamic agriculture went hand in hand with the progress and decay of Islamic science."[4]

Whatever the cause and effect relationships between the rise and decline of Islamic science the relevance of the rich historic past to the difficult tasks of the present is slight. The problem is to acquire the indigenous capabilities to keep pace with world scientific and technological progress. This is acknowledged by the revivalists such as Dr. Madkour who calls for "an era of progress and renewal", while praising the 40 plus universities in the Arab world for their "contributions to the expansion of knowledge through original research, especially on problems of particular relevance to their region's material and human resources."[5] It is not enough in the late 20th century to recall the scientific triumphs of the 12th century.

The debate over the relevance of Islam to the ability to do original science and technology is no more fruitful than the debate over the relevance of the past. The enormous diversity of the Islamic world with its 800 million believers, a majority of the population in over 40 countries, makes generalization hazardous. The evidence is convincing that profoundly Islamic Pakistan with a population nearing 100 million is building an important R & D capability.[6] No Arab state nor Turkey has achieved a comparable capability in absolute or relative terms.

The contribution of the contemporary Islamic world to global non-proprietary science and to proprietary technology can be measured by patents, copyrights, publications and citations in major scientific journals and by other means. It is limited but growing; small but not insignificant. Since most individual Islamic countries have only exercised independence for 10-40 years, and only began formulating science and technology policies in the 1960s, no definitive assessment of their R & D capabilities and achievements can yet be made. (Table 4.1)

Instead the argument over Islam and modern science is often phrased in terms of belief systems, secularization, and language. Essentially those who find a conflict between Islam and modern science maintain that as an absolute religion based on revealed truth that Islamic theology and pedagogy are antithetical to scientific inquiry. It is further argued that as a theocracy Islam tends to suppress inquiry, especially when it may threaten religious belief. Finally the argument concludes that the emphasis on the Koran, Arabic, and Koranic rote learning impedes the mastery of English or other languages needed to study science and technology. As a corollary Islamic attitudes and practices towards women are held to make their access to science and technology learning difficult.

These alleged Islamic barriers to modern science have been discussed extensively and polemically.[7] They do not appear to me to constitute the primary obstacles to the achievement of indigenous R & D capabilities. Turkey has been legally and culturally secular for more than 50 years with a romanized language that facilitates the learning of French or English yet its national science and technology capabilities remain weak. Egypt has been avidly importing western science and technology since the Napoleonic invasion and has formally educated thousands of scientists and engineers yet its capabilities are seriously deficient.[8]

TABLE 4.1
Distribution of Publishing Scientists in the Arab World by Country

	1967	1968	1969	1970	1971	1972	1973	1974	1975	1976
Algeria	22	27	28	46	52	60	62	49	69	63
Kuwait	2	1	3	2	9	16	39	42	40	56
Lebanon	58	67	67	89	93	109	106	174	172	100
Libya	3	3	2	5	12	16	16	19	28	31
Iraq	32	29	42	44	52	60	60	74	90	94
Jordan	1	1	5	7	5	7	7	22	13	15
Morocco	11	5	7	12	13	13	19	16	23	31
Syria	2	2	2	1	4	6	9	7	11	10
Saudi Arabia	8	12	6	14	14	14	17	31	36	57
UAR(Egypt)	293	295	348	443	442	547	433	648	738	731
Tunisia	3	13	13	17	22	20	20	29	37	39
Sudan	30	38	58	70	58	61	59	81	97	93
Yemen and others	0	0	0	0	2*	4**	0	0	1	3
Total	465	493	581	750	753	933	847	1,292	1,355	1,323
Israel	1,125	1,243	1,542	1,739	1,787	2,304	2,401	-----	2,935	3,291

* 1 from Yemen and 1 from the UAE

** 2 from Yemen, 1 from Oman and 1 from Somalia

Sources: for 1967 figures, 'Measuring the Size of Science' Derek J. de Solla Price, The Israel Academy of Sciences and Humanities, Proceedings IV, 6 (1969), p. 106; for 1968 onwards, International Directory of Research and Development Scientists (Institute for Scientific Information, Philadelphia, 1967, 1968, 1969, 1971). This publication is now superseded by: ISI's Who is Publishing in Science. The figures have been totaled from the index of authors in each country which is printed at the end of the volume. Reprinted by permission of St. Martin's Press, (c) 1980, from A. B. Zahlan, Science and Science Policy in the Arab World.

While Algeria has had a continuing post-independence conflict over choice of language (Arabic vs. French) for higher education, it has invested in intensive training in both French and English for its technocrats. Iran under the Khomeini regime has barred much but not all secular learning. However during the two decades of massive imports of S & T under the guise of cultural modernization national capabilities failed to take hold. Partial evidence comes from the current problems in maintaining and operating imported military equipment.

It is simply too easy to fault Islam as a religion and culture for the problems experienced in going from technology transfer to attainment of indigenous capabilities in much of the Middle East. Nor is it possible to provide meaningful empirical evidence for this charge. Publicly with massive rhetoric every Middle Eastern government from the People's Republic of South Yemen to the Royal Kingdom of Morocco proclaims its commitment to the development of science and technology.[9] In practice this usually means the search for technology transfers; an indication that the problems are in modern concepts rather than Islamic precepts.

No Home for Science and Technology

Science and technology can only function and grow where there are critical masses of researchers, facilities, flows of information, and sustained funding.[10] Results can only be produced when researchers have been trained, tests replicated, basic research done elsewhere absorbed, and continuity established. Twentieth century science-driven technology depends on large-scale interconnected bureaucracies unlike the one-man labs of 19th century researchers. Homes for research can be provided by universities, government ministries, public sector corporations, autonomous government labs, local private firms, foreign firms with local or regional branches, or various combinations and permutations. Even shop-floor or informal R & D depends on support systems and infrastructure, especially research incentives.

Science and technology in much of the Middle East has no secure home and can barely function. The lack of institutionalization prompts the most able researchers to emigrate, prevents the training of future researchers at home, and produces results that either are published abroad but find no audience at home, or else are published locally for non-existent audiences.

The dangers of generalizing for 18 countries with a total population over 270 million are clear. Fortunately there have been several comprehensive surveys of science and technology in the region since the 1960s that document the weak state of institutionalization, the brain-drains within and outside the region, and the poor linkages between researchers and potential users.[11] While there are exceptions and good research is being done in several countries on problems such as arid lands, dry lands farming, geology, and remote sensing the general patterns of deficient institutionalization continue to apply throughout the region.

Universities

Most universities in the Middle East are publicly controlled and funded and were established after World War II. Most are oriented towards undergraduate teaching with little time, support, or incentives for sustained team research. Where good research exists, as in Middle East Technical University in Istanbul, in engineering, and agronomy at the University of Morocco, and biology at the University of Jordan it is often through the efforts of a handful of individual faculty.

Governments fear political unrest from faculty and students and express their distrust of state universities in low funding, especially for long-term projects such as research. The highly politicized faculty and students are often uninterested in research which has few career returns. What research that occurs is sometimes basic and oriented at publication abroad and emigration, or else short-term applied work without an adequate data-base. The two principal private universities in the region, American University of Beirut, and American University of Cairo, are among the few institutions that have been able to find funds at home and internationally to keep research teams together. The Saudi Arabian government is alone in committing substantial research funds to its own universities which are still primarily staffed by foreigners. There are very few links between university researchers and any kinds of users. Governments generally prefer to import short-term consultants rather than rely on university faculty, and the private sector and external donors follow suit. Since university research does find users in Malaysia and Pakistan the evidence indicates that Islamic culture is not the problem. Rather it is the desire of Middle Eastern governments to prevent universities from acquiring independent power bases. For instance in Turkey during the last two decades government relations have deteriorated with the older traditional universities and their newer regional rivals. Governments in Egypt, Libya, Syria and elsewhere prefer to pay faculty and students as sinecures rather than run the risk of their developing contacts and ties off campus.

The most important manifestation of this research impotence is the lack of major graduate and post-doctoral research centers in the Middle East. This is partly due to the lack of regional university cooperation. The two principal research centers are internationally supported and operate outside universities. These are the International Institute for Theoretical Physics in Trieste, Italy, pioneered by Dr. Abdus Salam, the Pakistani Nobel prizewinner, and the International Centre for Agricultural Research in the Dry Areas (ICARDA) in Aleppo, Syria.

Writing in 1969 A. B. Zahlan, then Professor of Physics at American University of Beirut decried the "fundamental and universal assumption of all the Arab countries and of all the universities operating in these countries".."that the only source of advanced training should be at the institutions of advanced countries." This assumption continues to be prevalent throughout the Middle East. Moreover scholarships and trips to study abroad have become major government patronage devices, whether in Syria to the Soviet Union, or in Saudi Arabia to the U.S. and West Germany. The sending abroad annually of thousands of Middle Eastern students to do undergraduate and graduate work in the sciences and engineering reinforces the technology transfer concept and vitiates the need for local capabilities.[12] Study abroad has become a way of life rather than a means of permitting only the very ablest and most mature students to learn how to do research. Scattered evidence suggests that those who study abroad frequently come back as administrators rather than as researchers and lacking in the very abilities to improvise in the face of local scarcities that are most needed. It is understandable in Egypt and elsewhere that there is an acute shortage of competent locally trained lab technicians and that university degrees and possible study abroad have been established to provide incentives.

Government Ministries

Although some applied research is to be found in agriculture, mining, public health and other ministries in countries such as the Sudan, Morocco, Egypt, and Tunisia, it is generally short-term, low quality, and subject to high staff-turnover. Low salaries, lack of critical masses of researchers, facilities, career patterns, information channels and other obstacles plague ministry-based researchers. (Table 4.2) There is very little recognition of any significant role for research in the routine of these bureaucracies, even in agriculture where extension is often separated from research.[13] Applied public health research in Egypt and the Sudan to attack bilharzia and related diseases has been reliant on external funding and basic research for its advances.

Public Sector Corporations

The largest and most complex organizations in the Middle East aside from the military in terms of volume of sales and numbers of employees are the parastatals such as the state airlines and shipping corporations, the Moroccan government potash monopoly, and the state petroleum corporations and holding companies such as the Saudi Arabian Basic Investment Corporation (SABIC). Rarely do they have a formal research department, staff, or budget, and there is little evidence of informal shop-floor research. Instead they are mainly importers of transferred technology which has become a lucrative way of life. Often in Kuwait, Saudi Arabia, and the Gulf these are joint ventures which rely on foreign partners for technical and management staff, maintenance, and operations. Learning by doing cannot occur under these arrangements.

Parastatals are not users of local research. They shun local universities and engineering firms in their conviction that imported know-how is superior. There are exceptions such as the University of Petroleum and Minerals in Saudi Arabia. The most interesting exception are the two Algerian state petrochemical and electronics corporations. Rather than work with local universities they have established their own research institutes, trained several thousand students in English in the U.S. and attempted to develop in-house research capabilities until hit by falling oil prices. The general pattern though is for parastatals to regard research as "turf" to be jealously guarded rather than obtained through local sub-contracting.

TABLE 4.2

Estimate of Financial Resources Committed to R & D

Country	National expenditure on R & D In US$ (millions)		R & D as % of GNP		Per capita expenditure on R & D (US$)		Average annual expenditure per R & D scientist US$	
	1965	1976	1973	1976	1965	1973	1976	1973
Algeria	4-6*	2.6	0.13	0.02	0.3-0.5*	0.6	0.16	30,000
Bahrain	--	1.3	--	0.22	--	--	5.43	--
Egypt	--	85.5	0.83	0.89	--	2.2	2.37	7,235
Iraq	0.5-0.7*	90.1	0.25	0.65	0.07-0.10*	2.3	8.66	16,840
Jordan	0.05-0.17	6.5	0.31	0.52	0.026-0.09	1.1	2.50	15,390
Kuwait	0.25-0.5	45.8	0.01	0.30	0.6-1.2	0.8	53.25	17,100
		(1974)					(1974)	
Lebanon	1-2.7	5.7	0.40	0.17	0.43-1.2	2.4	1.89	22,400
Libya	--	48.3	--	0.36	--	--	21.00	--
Morocco	--	1.5	--	0.02	--	--	0.09	--
Oman	--	1.2	--	0.07	--	--	1.71	--
Qatar	--	1.0	--	0.04	--	--	11.12	--
Somalia	0.125**	4.2	--	--	--	--	--	--
Saudi Arabia	0.8-1.0	--	--	--	0.12-0.15	--	--	--
Sudan	2.5-4.5	8.7	0.33	0.21	1.2-0.35	0.5	0.5	31,176
Syria	--	7.3	--	0.14	--	--	1.08	--
		(1977)					(1977)	
Tunisia	2.6*	21.2	0.30	0.52	0.6*	1.1	3.79	10,900***
Yemen Arab Republic	--	2.9	0.25	--	--	0.3	--	35,270

*For 1966. **For 1967, UNESCO 31,p.96. ***A figure of $20,000 is quoted in UNESCO 31,p.87.

Sources: for 1965 figures, UNESCO 6, p. 11; for 1973 figures, UNESCO 38, p.21; for 1976 figures, the Kuwait Fund for Arab Economic Development, A Proposal for Establishing an Arab Fund for Scientific and Technological Development, Table 2, p.35, henceforth AFSTED Report and Hafez, R & D, Tables 4 and 8. Reprinted by permission of St. Martin's Press (c) 1980 from A. B. Zahlan, Science and Science Policy in the Arab World.

Government Research Centers

Several governments such as Egypt and Kuwait have established multi-purpose autonomous research institutes. Those in Egypt have been plagued by the problems of low salaries, part-time researchers, lack of users, sinecure posts, and lack of continuity.[14] Even the parastatal corporations choose not to work with these institutes. The Kuwait Institute for Scientific Research is well-funded but dependent on high turnover expatriate staff with a lack of research priorities.

The establishment of national science and technology councils and embryonic planning efforts in many countries since the 1960s has resulted in intense conflicts within small scientific and engineering communities. Frequently there is excessive centralization of R & D funds and priorities, often in new universities or centers. Jordan, Syria, and Tunisia have experienced persistent conflicts over funds and priorities. Handfuls of isolated, fragmented, and subcritical groups of researchers compete for the attention of the national science and technology councils or the minister or junior minister for science and technology. (Table 4.3)

The Saudi Arabian National Council for Science and Technology (SANCST) hired the U.S. National Science Foundation (NSF) to get started, thus importing an institutional model.[15] It has a research fund and a peer review system for its allocation and works primarily with the Saudi universities. Otherwise Saudi Arabia, like Kuwait and the other oil-exporters has subordinated R & D spending beneath several budgets for the import of capital goods. Local funds are allocated to training of personnel in the operation of imported technology but not in its adaptation. Government centers for desalination, solar energy, nuclear technology, and ecology concentrate on technology transfers rather than research in Saudi Arabia, Kuwait, and the Gulf states.

Multinational Corporations

Although multinational corporations have established R & D activities in many developing countries they generally do not do so in the Middle East. The reasons are several. There are few if any fiscal, monetary, labor cost, or other incentives to establish local or regional R & D. Major customers, especially oil-exporting governments, want state

TABLE 4.3
Science policymaking in selected countries

Country	Institution	Projects/programme	Staff
Algeria	CNRS	Science policy envisaged	Professional
Iraq	FSR, Science Policy Unit	Sectoral science and technology plans, overall plan envisaged	Professional, technical
Jordan	RSS	Initiatives taken for formulation of national science policy and plan	Professional, technical
Kuwait	KISR, Techno-economics Division	Long-term technological requirements studied	Professional, technical, foreign
Lebanon	NCSR	Science plan included in 1972-77 development plan	Professional, technical
Pakistan	Pakistan Science Foundation	Science policy developed after lengthy national debate	Professional, technical
Saudi Arabia	SANCST	National science and technology plan part of the third development plan (1981-85)	Professional, technical, foreign
Syria	SCS to be integrated in National Council for Science Policy	National science and technology plan envisaged	Professional, technical
Turkey	TUBITAK	Sectoral science and technology plans	Professional, technical

Reprinted by permission of Longman (c) 1982 from Ziauddin Sardar, *Science and Technology in the Middle East.*

of the art technologies rather than less-expensive and proven or used capital goods that might require local repair. This is particularly true of major weapons systems which are imported to be used and then discarded. Local universities produce few research-grade students, transferring foreign researchers is expensive, and use of trouble-shooters and consultants has become customary. The conflicts that have occurred with R & D giants such as IBM over demands for local research in Brazil, India, and elsewhere have no counterpart in the Middle East.[16] Rich and poor governments in the Middle East prefer to import the latest items rather than locally upgrade. The tendency to equate best with most expensive and complex is pervasive.

The potential for the multinationals initiating applied research on a regional or subregional basis exists. For example the Arabian American Oil Company (ARAMCO) began a demonstration irrigation project in Saudi Arabia in the 1940s and continued water research as did Shell in Libya. Currently oil-exporting governments show little interest in providing multinationals with local research incentives, preferring joint venture package projects relying on South Korean, Pakistani, Philippine, and other foreign temporary migrant workers who are required to live in project enclaves and do little training of local people.[17]

Local Private Sectors

There are large-scale engineering and contracting firms in Egypt, Lebanon, and Turkey which are winning design and construction contracts outside their own countries.[18] There are privately owned firms exporting manufactured goods within the region. In the oil-exporting countries, especially Saudi Arabia, there are many local sub-contractors who carry out a wide range of building and other activities. There is no obvious shortage of entrepreneurs, although not many are card-carrying M.B.A.'s. There are numerous smaller industries and enterprises with a potential for shop-floor learning.

Many of the locally owned private firms are family businesses, do not have research organizations or capabilities, and are highly dependent on government favors. It is difficult to see them as the forgers of an independent research capability or even substantial shop-floor learning unless and until governments provide suitable environments and incentives. Turkey is the mostly likely country in which this may happen since there is widespread disillusionment with the ability of the state research centers to disseminate research, and the parastatals to do competitive research. It could also happen in Egypt if the private sector had consistent and substantial incentives not to import technologies for all its needs. Like Pakistan, Turkish firms have become increasingly competitive in winning Middle Eastern contracts.

External Donors

Bilateral and multilateral donors in the Middle East are prone to urging the use of imported hardware, software, and consultants. Hard-pressed governments often accept technology transfer from donors as a means of acquiring foreign exchange. External funding has been hard to obtain for the local costs of research.

Non-Profit Foundations

One sees the possible relevance of the Islamic past in current efforts to promote science and technology through non-profit institutions modeled partly on the historical religious foundations and trusts (waqfs) used to alleviate poverty. Trusts have been established in Turkey for historical reconstruction and restoration, non-profit institutes for funding research have been created in Kuwait, and Saudi Arabia, The Islamic Foundation for Science and Technology for Development (IFSTAD) was chartered by 41 governments in 1979, and the Ismaili world-wide community supports the Aga Khan Fund for Islamic Architecture.[19] These are all significant initiatives and represent an attractive institutional model although one that does not build researcher-user ties.

The amounts of funds committed to research through these foundations are modest and miniscule compared to other forms of Middle Eastern non-profit spending. The well-endowed Development Banks and Islamic Funds established with oil revenues in Kuwait, Saudi Arabia, and elsewhere have invested in general-purpose soft loans and the building of mosques. The oil-exporters have poured billions into technology transfer at home and lent for this purpose abroad but not for research as such. As conservative traders and merchants they tend to be skeptical of investments in research and prefer to buy its results from others.

Intense efforts to create a regional non-profit

research funding vehicle have not yet succeeded. The Union of Arab National Research Councils (UANRC) was established in 1975 and includes Algeria, Egypt, Iraq, Jordan, Kuwait, Palestine, Sudan, and Tunisia and has lobbied for the proposed Arab Fund for Scientific and Technological Research. The Islamic Foundation has a $50 million initial plan to support research on Islamic values and ethics on science and technology, the brain drain, and related topics of interest to its Arab and non-Arab members.[20] In spite of these and other efforts must Middle East researchers remain dependent on their own governments or external donors for support. (Table 4.4).

The lack of institutionalized homes for science and technology impedes the incremental growth which is essential. The oil-exporters make major investments in new facilities such as universities or research centers and then as oil prices decline cutback on operating costs. Research whose returns cannot be predicted is the first to go, especially since these new hospitals, universities, and other centers have extensive day to day responsibilities. Where foreign exchange is scarce researchers are kept on the payroll as in Egypt but much research depends on the availability of external funding. This results in clashes over priorities and elaborate tug of wars to reconcile donor-researcher and host government interests.[21]

Linkages

Nowhere in the Middle East are there on-going effective linkages between researchers and research-users. It is the lack of linkages that perpetuates the dependence on imported technology and the failure to learn by doing. Faced with foreign exchange constraints shop floor R & D does occur but there are few information channels for its diffusion. Occasionally a new local technology such as the drip-method of irrigation developed in Israel will be rapidly informally diffused in neighboring countries. Historically local R & D played an important role in the cotton and textile industries of Egypt and the Sudan but no counterpart has developed in the modern urban manufacturing sectors. Government-funded industrial technology centers in Egypt have floundered as larger businesses prefer to import and smaller firms are wary of poorly trained government engineers. Again the primary obstacles to researcher-user linkages appear to be institutional rather than characteristics of Islamic society.

TABLE 4.4

Scientific Manpower in the Arab World

		TOTAL STOCK			
		Scientists and engineers		Technicians	
Country	Year	Total Number	Number per 100,000 pop.	Total Number	Number per 100,000 pop.
Algeria	--	--	--	--	--
Bahrain	1971	928	935	--	--
Egypt	1973	593,254	1,657	--	--
	(2)				
Iraq	1972	43,645	432	24,689	244
Jordan	1973	4,288	170	1,089	43
Kuwait	1973	10,754	1,139	2,930	310
Lebanon	1973	37,000	1,163	8,000	250
	(2)				
Libya	1973	8,319	392	10,602	500
Morocco	--	--	--	--	--
Oman	--	--	--	--	
Yemen, People's Dem. Rep. of	--	--	--	--	--
Qatar	1974	1,352	1,572	577	671
Saudi Arabia	1974	33,376	395	--	--
Somalia	--	--	--	--	--
Sudan	1972	13,792	84	2,639	16
Syria	--	--	--	--	--
Tunisia	1974	3,421	61	7,714	140
United Arab Emirates	--	--	--	--	--
Yeman Arab Republic	1975	1,394	22	680	11

(Continued)

TABLE 4.4 (Continuation)
Scientific Manpower in the Arab World

	ENGAGED IN R & D						
	Scientists and engineers			Technicians			
Country	Year	Total Number (FTE)**	Number per 100,000 pop.	Year*	Total Number*	Total Number	Number per 100,000 pop.
		(1)				(1)	
Algeria	1973	242	1.6	1966	48-60	100	0.6
Bahrain	--	--	--	--	--	--	--
Egypt	1973	10,655	29.8	--	--	--	--
Iraq	1974	1,486	14.1	1966	280	376	3.6
Jordan	1973	180	7.1	1965	51-100	41	1.6
Kuwait	1973	176	18.6	1965	100-210	15	1.6
Lebanon	1973	340	11.4	1965	140-280	255	7.4
Libya	1973	50^+	2.4	1970^s	70	142	6.7
		(3)					
Morocco	1970	253	1.6	1970^s	253	290	2.5
Oman	--	--	--	--	--	--	--
Yemen, People's Dem. Rep. of	--	--	--	--	--	--	--
Qatar	--	--	--	--	--	--	--
Saudi Arabia				1965	105-210	--	--
Somalia	1970^s	24	--	--	--	--	--
Sudan	1973	249	1.5	1965	210-300	--	--
Syria	--	--	--	--	--	--	--
Tunisia	1972	550	10.1	1966^s	80	552	10.1
United Arab Emirates	--	--	--	--	--	--	--
Yemen Arab Republic	1975	60	1.0	--	--	52	0.8

(1) Data relate to higher education sector only.
(2) Government employees only.
(3) Full-time plus part-time personnel.

*Estimate **FTE = full time equivalent

[s]*Science Policy Studies and Documents, No. 31, National Science Policies in Africa* (UNESCO, Paris 1974), p. 90, henceforth referred to as UNESCO 31. Libya: 73% are part-time researchers, Morocco: 10% are part-time researchers, a figure of 318 researchers is given for Tunisia (1970) and for the Sudan: 470, 53% of whom are part-time researchers.
Sources: UNESCO document SC-76/CASTARAB/Ref. 1.
Science Policy Studies and Documents, No. 6, Structural and Operational Scheme of National Science Policy, Conclusions and Recommendations of the Meeting on Science Policy and Research Organisation in the Countries of North Africa and the Middle East, Algiers, 20-26 September 1966 (UNESCO, Paris, 1967) Table 11, p. 10, henceforth referred to as UNESCO 6.
Reprinted by permission of St. Martin's Press from A. B. Zahlan, *Science and Science Policy in the Arab World*, (c) 1980.

Regional Cooperation

The advantages of regional cooperation is science and technology have been advocated by the Arab League, UNESCO, and other international organizations.[22] The Arab League Educational, Cultural, and Scientific Organization (ALECSO) created in 1970 has made important progress in providing Modern Arabic with a scientific and technical vocabulary, and improvements in texts, typewriters, telex machines, journals, information exchanges, and computer materials. The Gulf States share a new sub-regional university and an education research center. There are frail regional and subregional associations of professionals in the health sciences, chemistry, physics, agriculture, engineering, and other fields. Impressive work has been achieved on the teaching of English for science to Arabic-speakers in Saudi Arabia and diffused elsewhere. The Arab League has attempted to standardize curriculum and to improve elementary and secondary school science education producing supplmentary materials. Regional cooperation was given a further boost in 1985 with the launching by a Saudi astronaut in a U.S. spacecraft of the first Arab communications satellite jointly owned and operated by several governments headed by Saudi Arabia.

Progress towards sharing of research costs and tasks has been much slower. The International Center for Agricultural Research in Dry Areas in Syria is the only Middle East member of the International Agricultural Research Center Consortium. UNESCO sponosored a major meeting of Middle East Ministers responsible for science and technology (CASTARAB) in 1976 which generated proposals for regional communications satellites, designated research universities, and research priorities for arid lands, ophthalmology, and Islamic medicine.[23] Political and economic difficulties have blocked most of these proposals, as well as funding for the Islamic Foundation.

The case for regional cooperation and formation of networks of researchers remains strong in spite of the frustrations to date. Egypt, Turkey, and Iran are the only countries with the human and financial resources to achieve even moderate S & T capabilities by world standards. The others lack funds and researchers to do more than select and pursue a few research priorities. This holds true even for the oil exporters who can buy research but not good researchers from abroad.

The preference has been for bilateral cooperation in S & T with non-regional governments such as the US, USSR,

France, West Germany, Japan, and China. This resulted in a proliferation of bilateral agreements during the oil-crisis years of the 1970s. However many of these agreements were seen by the non-regional partner as a vehicle for selling technology. There can be little cooperation in research where indigenous research capabilities are so limited.[24] While the Soviet Union and individual Eastern European governments continued to sign such agreements in the 1980s they had little competitive technology of a non-military nature to offer.

The Future

The oil-exporters are mostly pursuing strategies of increasing the value added of their petroleum and gas exports through imported technologies which provide transformation. Hence the petrochemical plants, gas liquefacation processes, and even the iron and steel mills and aluminum refining plants using cheap gas. This capital-intensive approach is intended to produce a new generation of exports. As long as these joint ventures pressure downstream clients to accept these products as a condition of buying oil and/or sell these products at discounts, markets should be available. However profit margins may decline and some new plants in the Gulf and elsewhere may require heavy subsidies.

These strategies will in no way increase significantly indigenous R & D capabilities. Not unless drastic changes are made in the joint ventures. Host-country nationals are learning far more about foreign stock exchanges and money markets than any learning by doing from operation of these massive investments. Continued reliance on well-paid and exotic foreigners to run these plants will be a source of persistent conflict and frustration, especially if exports have to be subsidized. Little will have been done to prepare these societies for the time when non-renewable resources are seriously depleted. A few people will note that it is the Pakistanis and South Koreans that will come out ahead through achieving and exporting R & D rather than the pious wealthy. They will have both been paid handsomely to design and to build the petrochemical plants and in doing so will have learned how to build even better ones at home.

The stragglers who do not have petro-dollars will have to pursue their own slow, hard struggle to master science and technology. They will need to develop some competitive manufactured exports within the region, as well as to continue to export brainpower to their neighbors. They will have to acquire indigenous agricultural capabilities or risk becoming chronic food importers.

Agricultural research has to be the priority for Morocco, Syria, Sudan, Tunisia, Yemen, and other countries.[25] It is the one technology which is least transferable from abroad due to the uniqueness of soil, climate and other growing conditions. The vast sums being spent by Saudi Arabia and Libya on irrigating the desert with underground water are beyond the means of others and ecologically questionable. Underspending on agricultural research is endemic in the region and a certain way of compromising the future as Algeria has already learned. The temptation of massive transferred agricultural and industrial projects outdazzles the hard road of increasing agricultural productivity. External donors are asked to become food suppliers as in Egypt and Tunisia. The prospect of a dozen poverty-stricken Middle Eastern states permanently importing food on a non-commercial basis is highly undersirable. Critical, sustained agricultural research and extension can increase yields and head-off import dependence. It is the national, regional, and international donor imperative.

NOTES

1. Nathaniel C. Nash, "What's New in Israeli High Tech," New York Times, August 12, 1984.

2. Nathan Rosenberg, Inside the Black Box: Technology and Economics (Cambridge, Mass.: MIT Press, 1982).

3. Ibrahim Madkour, "Past, Present, and Future," in John S. Badeau, ed., The Genius of Arab Civilization, Source of Renaissance (Cambridge, Mass.: MIT Press, 1983), 2nd. ed. p. 243.

4. Ibid. Further discussions of the history of Islamic science are found in Marshall G. S. Hodgson, The Venture of Islam, Vol. 2, The Expansion of Islam in the Middle Period (Chicago: University of Chicago, 1974), pp. 165-174; Andrew M. Watson, Agricultural Innovation in the Early Islamic Period (New York: Cambridge, 1983), p. 145; Nancy Elizabeth Gallagher, Medicine and Power in Tunisia, 1700-1900 (New York: Cambridge, 1983), p. 7-13; and Ziauddin Sardar, ed., Science and Technology in the Middle East (New York: Longman, 1982), "Islamic Science", pp. 18-23.

5. John S. Badeau, ed., _The Genius of Arab Civilization_ (Cambridge, Mass.: MIT Press, 1983), p. 245.

6. A. B. Zahlan, _Science and Science Policy in the Arab World_ (London: St. Martin's, 1980); Ziauddin Sardar, _Science and Technology in the Middle East_ (New York: Longman, 1982), "Pakistan", pp. 184-218.

7. John S. Badeau, op. cit. "A Guide for Further Reading," pp. 247-250; Claire Nader, A. B. Zahlan, eds., _Science and Technology in Developing Countries_ (London: Cambridge University Press, 1969), p. 263. English is the language of instruction at the University of Petroleum and Minerals in Saudi Arabia and at the American Universities at Beirut and Cairo. It is a second language at universities throughout the Middle East except North Africa where French remains the first or second language of university instruction.

8. On Egypt see Clement Henry Moore, _Images of Development, Egyptian Engineers in Search of Industry_ (Cambridge, Mass.: MIT Press, 1980). On Turkey see A. Kemal Ozinonu, "Pattern of Scientific Development in Turkey, 1933-1966," pp. 141-173, and Osman Okyar, "The University and Regional Development," pp. 367-400 in Claire Nader, A. B. Zahlan, eds., _Science and Technology in Developing Countries_ (London: Cambridge University Press, 1969).

9. Ziauddin Sardar, ed., _Science and Technology in the Middle East_ (New York: Longman, 1982) contains a country by country survey of government science and technology policies.

10. World Bank, _Agricultural Research, Sector Policy Paper_, Washington, DC, June 1981 contains an excellent summary of the basic components of agricultural research and global data. On Middle East agricultural research and extension services see Afif I. Tannous, "Organizing Science and Technology for Agricultural Development," in Claire Nader, A. B. Zahlan, eds., op. cit. pp. 61-84.

11. Comprehensive surveys of science and technology in the Middle East have been undertaken since the 1960s using a variety of methodologies and emphases. Taken together they provide a useful perspective on the evolution of problems and policies. Useful on North Africa is C.R.E.S.M. _Politiques Scientifiques et Technologiques au Maghreb et au Proche-Orient_ (Paris: CNRS, 1982). Valuable historically is Claire Nader, A. B. Zahlan, eds., op. cit. drawing on a November 1967 Conference at American University in Beirut with reports from a number of countries. Other regional surveys include: UNESCO, _Science and Technology in the Development of the Arab States_, Science Policy Series No. 4, Paris, 1972; UNESCO, _National Science and Technology_

Policies in the States: Present Situation and Future Outlook, Science Policy Series No. 38, Paris, 1976; J. Davidson Frame, Indicators of Science and Technology Efforts in the Middle East and North Africa, U.S. Agency for International Development, Washington, DC, December 1978; A. B. Zahlan, Science and Science Policy in the Arab World (London: St. Martin's, 1980); and Ziauddin Sardar, ed. Science and Technology in the Middle East (London: St. Martin's, 1982).

12. The status of universities in the Middle East is discussed in two thoughtful essays in Claire Nader, A. B. Zahlan, eds. op. cit. See Matta Akrawi, "The University and Government in the Middle East," pp. 335-362; and A. B. Zahlan, "Problems of Educational Manpower and Institutional Development," pp. 301-334. The quotation is from the essay by Zahlan, p. 303. The total number of Middle Eastern students abroad at undergraduate and graduate levels in North America, Eastern and Western Europe, and the Soviet Union may be close to 200,000 during the 1980s. An Institute of International Education survey reported 60,660 enrolled in the U.S. in 1983-84; down 9.8 percent from the previous year. Chronicle of Higher Education, September 5, 1984, p. 21. Figures by nationality studying in the U.S. in in 1983-84 were Iran-20,360, Saudi Arabia-8,630, Jordan-6,890, Lebanon-6,680, Kuwait-3,810, Israel-2,610, Egypt-2,340, Syria-1,940, Iraq-1,730, Libya-1,710 and United Arab Emirates-1,260. A. B. Zahlan estimated in the mid 1960s that 25,000 Arab students were abroad, two-thirds as undergraduates, with 7,000 in the U.S. A. B. Zahlan, C. Nader, op. cit. p. 306.

13. World Bank, Agricultural Research, Sector Policy Paper, Washington, DC, June 1981.

14. Clement Henry Moore, Images of Development, Egyptian Engineers in Search of Industry, op. cit. pp. 84-109.

15. National Science Foundation, Science Indicators 1982, Washington, DC, 1983.

16. The Saudi Arabian Basic Investment Corporation (SABIC) has been considering the possibility of joint ventures to promote local production of high technology items without local research. Contrast this with the Brazilian experience as analyzed in Paulo Bastos Tigre, Technology and Competition in the Brazilian Computer Industry (New York: St. Martin's, 1983).

17. J. S. Birks, C. A. Sinclair, International Migration and Development in the Arab Region (Geneva: International Labour Office, 1980) is a country by country discussion of Arab and non-Arab migrant workers in the

Middle East.

18. Louis T. Wells, Jr., Third World Multinationals, The Rise of Foreign Investment from Developing Countries (Cambridge, Mass.: MIT Press, 1983) includes data on Third World multinational companies operating in the Middle East. Studies of shop-floor engineering and technology adaptation practices by Middle Eastern firms need to be conducted.

19. The Aga Khan prizes for Islamic Architecture are described in Renata Hood, ed., Architecture and Community: Building in the Islamic World Today (Millerton, New York: Aperture, 1983). These prizes are awarded to Islamic and non-Islamic firms and architects for contemporary architecture, normally non-religious.

20. Regional cooperation in support of science and technology is discussed in Ziauddin Sardar, op. cit. pp. 90-119. The activities of the Arab League, the Conference of Arab Ministers Responsible for the Application of Science and Technology to Development, the Islamic Foundation for Science and Technology for Development, and the Union of Arab National Research Councils are outlined.

21. During the late 1970s Egyptian researchers were interested in obtaining ultra-sound equipment. Through the Egypt-U.S. National Science Foundation agreement funded by U.S. AID the Egyptians were able to obtain ultrasound equipment for applied biomedical research although this was strictly peripheral to the agreement and the basic research objectives of NSF.

22. See Ziauddin Sardar, op. cit. pp. 90-119 for a summary of efforts at regional cooperation. Also pp. 64-67.

23. UNESCO, Science and Technology Policies in the Arab States: Present Situation and Future Outlook, Science Policy Series No. 38, Paris, 1976. The 1985 launching of an Arab Communications satellite with a Saudi astronaut and a multi-government owned corporate structure based in Saudi Arabia was an indirect outcome of the 1976 CASTARAB meeting and proposals.

24. The U. S. National Science Foundation (NSF) is committed by statute to funding basic research and has found it difficult to locate viable partners in the Middle East. No cooperative research agreements were reached with Algeria or Turkey and limited agreements with Morocco and Tunisia were not renewed. AID funds were used to support a mostly applied research program with Egypt. NSF jointly funded with Kuwait and Saudi Arabia conferences and technical assistance. Only with Israel has a working bilateral cooperative research agreement been possible.

25. World Bank, Agricultural Research, Sector Policy Paper, Washington, DC, June 1981, pp. 38-47.

5

Africa: Frustration and Failure

Aaron Segal

More than two decades after attaining political independence most of Africa's fifty-two independent states have not significantly improved their indigenous science and technology capabilities. Indeed in many countries there is less research and fewer qualified researchers than there were prior to independence. Whether defined in terms of formal research institutes or less informal attempts to build intermediate and appropriate technologies, the track record of twenty-five post-independence years is dismal. The failure to establish indigenous capabilities has been an important factor contributing to low agricultural productivity, negative economic growth in many countries, the persistence of endemic health and environmental problems, and the inability to effectively transfer technology. Much of Africa is not involved in learning by doing, applied research, reverse engineering, or *in situ* science. It remains dependent on expatriate researchers and foreign research for its ideas, instrumentation, and all too often for application.

This critical evaluation is of course subject to qualifications. Four of the thirteen international agricultural research centers are located in Africa (Livestock in Ethiopia, Tropical Agriculture in Nigeria, Semi-Arid Tropics (Upper Volta sub-station), and Wheat and Maize (Zaire and Tanzania sub-stations). Important and valuable research is being conducted at these international research centers located in Africa and at other international research centers in insect physiology, desert locust control, onchocerciasis and other subjects.[1] Much of this research though is foreign funded and conducted. The lack of national indigenous research capabilities is a major obstacle to the effective dissemination and adaptation of the work being done at the international centers. The networks often run from the international research centers in Africa to the international scientific communities in their specialities.

The contrast between South Africa (SA) and the rest of Africa is enormous. South Africa has about 8,000 full-time researchers, more than the rest of Africa combined including Egypt and North Africa.[2] Its public and private sector expenditures on research and development (R & D) have been about $1 billion a year since 1980 and substantially more than the rest of Africa combined (Table 5.1). Tragically two-thirds of its R & D is spent on weapons and nuclear energy and only five percent on the agricultural and health needs of its people. Moreover almost all its researchers are white and the failure to invest in science education denies to the majority of its people a chance to study science or engineering. Science policy in South Africa and resource allocations are determined by a thirty member Science Advisory Council representative of the public, private, and university sectors and responsible to the President. It is the only science policy decision-making mechanism in Africa that can be said to produce results. Elsewhere science and technology policy institutions are weak, scattered between government agencies, lack meaningful non-government participation, and are generally ineffectual.[3]

Although by global standards South Africa is a middle-ranking science and technology (S & T) power, with a capability somewhere between that of Belgium and New Zealand, it is a science and technology colossus by African standards. Its meteorological, veterinary, geological and other services are relied on by most of Southern Africa which lacks its own capabilities. The story of the evolution of South African science and technology has been discussed elsewhere as well as its current status.[4] Its most acute problem is to replace an aging stock of all-white researchers in a society which has underinvested systematically in education. Meanwhile South Africa remains highly dependent on imported commercial technology and non-propietary science. It is capable though of a high degree of technological self-reliance, and very sophisticated end runs to counteract possible sanctions. It has had extensive experience with technological learning by doing, especially adaptive research with weapons systems.

South Africa has been successful to date in establishing indigenous capabilities, mobilizing these to defend white supremacy, and using S & T for objectives other

TABLE 5.1
Research and Development Expenditures for Selected African Countries

Country	Year	Expenditure in 1970 $ US	Expenditure as % of GDP	Expenditure Per Capita 1970 $ US
Egypt	1973	62710 mn.	0.82	2.12
Kenya	1971	13272 mn.	0.81	1.23
Zambia	1972	8072 mn.	0.50	1.96
Sudan	1973	7170 mn.	0.34	0.55
Cameroon	1970	6355 mn.	0.61	1.09
Ivory Coast	1970	5045 mn.	0.35	1.17
Mauritius	1977	2344 mn.	0.37	3.13
Upper Volta	1970	1486 mn.	0.38	0.28

<u>Source</u>: UNESCO, Statistical Yearbook 1978-1979, Paris, 1980

than economic or social development. The rest of Africa by contrast has failed or not tried to establish indigenous capabilities, and has been unable to mobilize S & T for any objectives.

Science and technology in Africa has had the support of rivers of rhetoric. The 1964 Lagos Ministerial Meeting organized by UNESCO established a series of indicative goals for scientific and technical manpower by 1980, none of which came close to being met. A series of OAU and UNESCO technical meetings preceded a major 1970 African science and technology meeting which resulted in a detailed survey of national research institutions and their capabilities. Subsequently, a 1974 UNESCO meeting at Ministerial level examined science and technology policies and recommended a number of national, regional, and all-African measures, few of which saw the light of day. Again a number of technical meetings preceded the 1979 UN Conference on Science, Technology, and Development in Vienna. The African contribution to, and benefit from, the Vienna conference was minimal.

The Myth Of Africa's Wealth

There are many reasons why science and technology have yet to mature and take hold in Africa. One of the most important is the persistence of the myth that Africa is wealthy in natural resources, that it has been exploited to its detriment for hundreds of years by rapacious foreigners, and that it needs only to recover and retain its wealth in order to enjoy prosperity. Nothing could be further from the truth. Out of 52 African countries, only five have known appreciable reserves of non-renewable oil or natural gas; only 14 have known significant deposits of valuable minerals; and most importantly, arable, well-watered land is in scarce supply throughout Africa.[5] The discovery of additional energy or mineral deposits, the improvement of agricultural or livestock yields, and the creation of additional arable land through irrigation all require the systematic application of science and technology; a science and technology at present available only from outside Africa (or SA) and often available on harsh terms.

Of all the gaps that separate Africa from the rest of the world the science and technology gap is probably the most critical, and the most profound. Ironically, the working paper for the 1974 UNESCO Ministerial conference declares that "in terms of land suitable for agriculture, existence of vast forests, potential for cattle-raising and fisheries, energetic resources and, last but not least, mineral deposits, Africa is on the whole a very wealthy continent."[6] There is no mention of human resources and no recognition that the science and technology needed to develop "a very wealthy continent" is far different from that of a very poor continent.

If one accepts that Africa has limited and highly unevenly distributed natural resources then the need for, and the kinds, of science and technology acquire a different perspective. Africa needs to be able to inventory and uncover its natural resources so as to expand them whenever possible. This suggests remote sensing, botany, natural products chemistry, hydrology, solar energy and other relatively low-cost technologies and disciplines. A second imperative, stemming from limited resources, is that research and development often need to be carried out in situ since African environments and circumstances cannot be readily replicated abroad. The need for in situ field research is particularly true in agriculture, forestry, geology, the earth sciences, tropical medicine and other fields. Finally, the R & D imperative must be to increase productivity, whether yield per acre or output per man-hour. Africa is simply too poor to engage in expensive and time-consuming exercises in large-scale economic redistribution. The economic pie is so small that redistribution can only spread the poverty.

The Colonial Heritage

Throughout Africa's pre-colonial history there was striking, if uneven, technological development.[7] Iron age and stone age societies often coexisted in close proximity. Agricultural, metallurgical, mining and other technologies were widely yet unevenly disseminated over both space and time. While there was much rich cosmological, theological, numerological and other speculation there was little pre-colonial science.

A century or more of colonial contact and rule brought about an embryonic but highly distorted science and technology infrastructure. Primary emphasis was often placed on applied research to foster export crops such as cotton, coffee, cocoa and rubber. Such research in the French colonies was conducted by institutes headquartered in Paris, with scores of field stations in the colonies. Universities were late to be established, often favoured law and the humanities, and were organized on expensive elitist lines aimed to train civil servants and lawyers rather than to generate research. Government research budgets were highly fragmented, unstable and directed at resolution of immediate problems. Little provision was made for technical education and service and maintenance of imported scientific equipment was sadly neglected.

Although prestigious colonial scientific institutes and societies existed in Brussels, London, Paris and pre-World War I Berlin, their African counterparts were feeble and expatriate-dominated; science and technology policy *per se* languished. The sudden rush towards independence in the 1960s meant that totally inadequate timetables for the training of African scientists could not be implemented. Thus, at the time of independence in most African states, a majority of working researchers consisted of expatriates, often on short-term contracts. Except at the University of Ibadan in Nigeria (founded in 1948) and a few other places, Africans seeking graduate education in the sciences and engineering had to leave the continent. The thousands studying abroad, even it they all returned home promptly, were less than enough to replace the turnover of expatriates.

The Post-Colonial Experience

The failures, disillusionment and disappointments of the post-independence years are often explained by shortages of manpower aggravated by brain-drains of African scientists, inadequate government funding and attention, and insufficient external aid. Although these factors are significant, they do not constitute an explanation for what has and has not happened, nor is it a useful response to grave problems. More local and foreign money producing more African scientists and engineers will not put right what is wrong, especially since the defective colonial infrastructure is still intact in most countries. A more analytical and specifically prescriptive approach to science and technology in Africa is needed. For instance, the most important problem is not the acute shortage of skilled manpower. The formal and informal education and training of a competent scientist or engineer entails 5-10 years beyond secondary school, with 3-5 years for laboratory and other technicians. There are few short-cuts if quality is the goal, and any rapid expansion of numbers requires increasing the pool of secondary school students with science and technology backgrounds, including mathematics and some laboratory or fieldwork. This, in turn, means addressing the lowest priority in African education: the millions of primary school-children miseducated in overcrowded classrooms by underpaid and unqualified teachers with little or no equipment. If less than 1% of African school-children entering primary school go on to study post-secondary science, engineering or technical subjects it is because during their primary education they were usually taught by rote, left ignorant and terrified of maths, and totally deprived of any introduction to biology, nature or science. It is at this level that the real wastage is occurring, and it is repeated in the extraordinarily high numbers of failures at secondary school science and mathematics examinations, and the disinclination to study these subjects at university level or beyond. The talent pool of researchers 10-20 years from now can only be expanded by dramatic improvements in primary education (Table 5.2).

The Need For Improved Education

A series of low-cost, small-scale experiments in teaching science in primary and junior secondary schools provides evidence of what can be done.[8] These experiments usually involve in-service training and incentives for teachers, preparation of low-cost materials for students based on environments with with they are familiar, revision of often archaic examinations, and constant monitoring and motivation. The British Council has been involved in a junior secondary school (ages 12-15) project in Botswana; a related effort was attempted in Ethiopia; the Ghana Association of Science Teachers has taken the initiative in that country; Kenya's Ministry of Education has promoted a new primary school mathematics course, as have Nigeria and Sierra Leone for science. Based in Ghana, the 11-country Science Education Programme for Africa (SEPA), founded by the exiled South African physicist, Hubert Dyasi, has concentrated on developing new curriculum materials and providing workshops for teachers. The Ivory Coast has pioneered the controversial use of educational television at primary school level on a national basis, while other Francophone countries have experimented locally with TV. Although there is no one formula for guaranteed success in improving pre-secondary science education, there is no substitute for experimental efforts.

Several thousands of African researchers who have chosen to remain in Western Europe and North America cannot

TABLE 5.2
The Social Returns to Education in Africa

Education level	Rate of return (percent)
Primary	29
Secondary	17
Higher	12

Source: George Psacharopoulos, "Returns to Education: an Updated International Comparison," in Timothy King (ed.), "Education and Income," World Bank Staff Working Paper, No. 402 (Washington, D.C., 1980).

begin to fill present shortages or future needs. Many are political exiles from Ethiopia and South Africa; others are unwilling to change their professional careers to a setting in which research is well-nigh impossible. India has found that the brain-drain can be arrested only by providing salaries and research incentives comparable to what is available elsewhere: Nigerian and Ghanaian experience supports Indias.

Placing priority on technifying primary schools obviously entails extensive long-range planning to provide places for substantially increased numbers of future science, engineering and technology students. Except for Egypt, where low admissions standards and guaranteed jobs for all university graduates has resulted in high-enrolment universities and over 400,000 post-secondary students, most African universities have enrolments of 5,000 students or less; few or no evening and adult extension facilities; excessively expensive boarding and residential requirements; and under-enrolments and under-utilization of science, engineering, and even some medical faculties. The build-up should concentrate on non-boarding technical schools with evening and weekend instruction, rather than more under-graduate universities which provide education rather than research. The high costs of post-graduate education in the sciences require that it be provided through designation of regional centres of excellence at existing African institutions. The African Association of Universities and its affiliated Deans of Science Faculties has begun the thinking which could enable Africa to offer high quality advanced education better than the sometimes dysfunctional education received now by sending students overseas.

The Need For Applied Research

While the supply problems of African science and technology often receive most attention, next to primary schools, the second priority problem is the lack of demand for research. Unless and until there is a constant demand input, and ongoing linkages between users of research and its providers, research in Africa will remain under-funded, unappreciated and largely unutilized. During the colonial period a weak but steady source of demand for research came from plantations and other agricultural, mining, and forest exporters, and from missionary societies for work on leprosy and other diseases. Most of the population had little awareness or understanding of research, and no ability to seek its aid. Instead, they were sometimes the objects of study by anthropologists and others. Little has changed during the post-independence period.

The foreign-owned private sector relies in Africa on proven technology, seeks to maximize returns in short periods and outside SA, shows little interest in even applied research. Nor are there fiscal or other government incentives for private sector R & D - unlike India, Brazil and some other developing countries. The African-owned private sector is almost desperately in need of engineering and other forms of research; but African universities have neglected business education and the problems of small businessess - so have engineering faculties. The handful of polytechnics are not organized for research; in some countries they are not allowed to accept private contracts. Yet links between the private sector, government and academic research institutes are a categorical imperative. Students should be expected to do research internships with local firms; credit made available for small-scale R & D; tax incentives offered; and other means sought to promote such linkages. A few examples - such as the University of Science and Technology in Kumasi and the University of Nsukka - show the potential benefit of cooperation between academe and the private sector.

The Failure Of Parastatals

Throughout Africa, governments have acquired large-scale parastatal enterprises [9], whose revenues rival those of multi-nationals active in their countries; yet they totally neglect R & D, both in their budgets and their staffs. The outstanding exception is Algeria, where the State petroleum corporation has established major petro-chemical and electrical-mechanical research institutes completely separate from the universities.[10]

In SA nearly 30% of total R & D funds come from the parastatal and other public sector enterprises. There is no comparable contribution anywhere else in Africa, even though parastatal enterprises are riddled with problems: engineering problems of adapting equipment to tropical conditions, surveying, geological and topographical problems, among many others. Parastatals usually hire foreign consulting firms to provide their research needs rather than contract with local universities or research institutes.

Any parastatal firm in Africa with an annual turnover of $100m should, broadly-speaking, have a permanent research capability; they could profitably double or triple presently miniscule national R & D budgets and, for the first time, set African science and technology on a post-colonial foundation. Experimental contracts between parastatals and local universities might be one way of initiating such links.

The Weaknesses Of Government Research

While ministries in most African governments - such as agriculture, health, industry, mines and education - have their own research units, much government research is like a sieve. They generally have two or three scientists and one technician, whereas the critical mass considered minimally necessary for multi-disciplinary research is ten researchers and four to five technicians (Table 5.3). Libraries are pathetically small, laboratories ill-equipped and out of date, spare parts unobtainable and, worst of all, they have few scientists with whom to share ideas. Research projects are invariably applied and short-term, selected arbitrarily with no coherent rationale or systematic funding. Work is intermittent since staff have other duties, often goes unpublished and unrecognized; there is no sense of progress being made. Expatriates provided by foreign aid or coming on short-term contracts make little contribution towards the status of local, under-appreciated researchers. Pay and working conditions are often distinctly inferior to those of universities, especially in such fields as agriculture and veterinary science. Civil service rules, dating to colonial times, fail to provide career ladders for scientists and recent graduates, who quickly switch to administration, private commerce or politics.

One fundamental problem of government R & D in Africa is its lack of demand. Extension services are barriers rather than transmitters of communication with farmers. How often does one see across the road from trial plots at African agricultural research stations, farmers cultivating in traditional ways?

Some rough and ready rules are needed. First, all research units with fewer than ten full-time research staff, plus technicians, could be disbanded at considerable savings. Similarly, if money is lacking for a minimum critical mass laboratory and library, it should not be wasted: it is better to contract with the local university, a consulting firm or an international organization than to throw away money into demoralizing, inferior efforts.

TABLE 5.3
Scientists and Engineers in Research and Development in Selected African Countries

Country	Year	Number of Research Scientists and Engineers	Research Scientists and Engineers Per 10,000 Persons
Ghana	1976	4084	4.0
Sudan	1974	3324	2.2
Tunisia	1972	818	1.5
Togo	1976	261	1.1
Senegal	1972	392	1.0
Ivory Coast	1970	319	0.7
Nigeria	1970-1971	2083	0.4
Kenya	1975	361	0.3

Source: UNESCO, Statistical Yearbook, 1978-1979, Paris,1980.

Second, all applied research should be the direct result of a demand from below - whether from farmers, villagers or manufacturers. It should be funded on a three to five year basis, a reasonable period to expect results. Rapid staff turnover is one of the weaknesses of African R & D. Third, linkages are needed between government applied research and more basic university work, as well as with private sector suppliers of fertilizers, pesticides, agricultural equipment and other materials.[11]

The Isolated Scientist

Isolation is the principal enemy of the African researcher. Cut off from a scientific community; lacking periodicals, scientific meetings, computers and even reprints, he risks rapid obsolescence. Numerous studies have shown that the lack of an indigenous research environment is the single most important factor contributing to the African brain drain.[12] Facilitating such an environment is a difficult and sensitive job. For instance, in Egypt where there are 4,000 or more researchers they do little or no research, publish - when they publish - in unknown local journals, often fail to keep up in their specialized fields, hold down several jobs at the same time, and spend little time in professional contact with their colleagues.[13] The establishment of national science societies in Nigeria, Ghana, Kenya and elsewhere, and the West African Science Association - which held its 12th regional meeting in 1980 - are excellent examples of steps towards creating a scientific community. There is also encouragement in the founding of the Council of Economic and Social Research Institutes, and the regional and all-African associations of psychiatrists, mathematicians, engineers, and other professionals. UNESCO, through its Nairobi science and technology office, has encouraged the emergence of an African scientific community, while other international organizations have also helped.[14]

Inevitably, though, the major work must go on at the national level through science education programs, development of publications at all levels of scientific literacy, science journalists in the media, use of radio and TV on a regular basis, and the provision of basic, functional schools, urban and rural, libraries and national centres for scientific and technical information. An experimental grassroots library programme in Mauritius illustrates low-cost methods of widely disseminating such information.[15] And, of course, one needs role models - comic-strips and popular magazines featuring African scientists, as glamorous as athletes, entertainers, and politicians.

A research environment may come more easily than research institutions. The nearly 60 African universities outside SA are under growing pressure to become 'nationally relevant' at the risk of losing what is left of their autonomy if they fail. The production of internationally-respected quality research is often at odds with what politicans mean by 'relevant'.

What is needed are a dozen or so regional centres of excellence at selected African universities with, perhaps, an emphasis on fields of present overall weakness such as business, hydrology, accounting, computer sciences and mining engineering as well as physics, biochemistry, chemistry, etc. These regional centres, like their counterparts in Latin America, could offer MA, PhD and post-doctoral work (the latter currently unavailable in the continent) to both African and overseas students. Other universities will need to develop a critical mass in the research fields in which they have special advantages, e.g. Khartoum in the field of archaeology.

Efforts at institutionalization should probably be directed at the new technical schools, rather than at the heavily bureaucratic universities. New courses for science and engineering technicians, new links with the private sector for in-service training, applied research and developmental projects funded by parastatal firms and government departments, need to be tried at this level.

Millions of Africans, burning with the desire to learn, cannot be accommodated in the universities or by university education, as presently constituted. It is this demand - tied to the need to improvise machinery at the shop-floor level, and the real shortages of technicians at every level- that should force the evolution of responsive institutions. External donors interested in African universities should examine these new opportunities.[16]

The international research centres established in Africa - such as the Agricultural Centre (IITA) at Ibadan; the Animal Disease Centre (ILRAD) at Nairobi; the Insect Physiology Centre (ICIPE) at Nairobi; and the Livestock Centre (ILCA) at Addis Ababa - are all impressive and costly transplants. They merit continued support, but it is important to nurture Africa's indigenous technical institutions.

Transfer Of Technology

Indeed, there is a danger of too much attention being paid to the glamorous possibilities of international, Pan-African and regional scientific and technical cooperation; to the far-flung branches of a tree which still lacks national roots.[17] For instance, the UN Economic Commission for Africa and the OAU mounted a major effort to involve Africa in the 1979 Vienna Conference. Yet the African position papers and views were lost in the confrontation between the Group of 77 developing countries and the developed countries over technology transfer - a burning issue for Brazil, India and other industrial exporters, but of only minor significance for most of Africa. Important modest efforts to organize new R & D, such as the WHO programme for tropical diseases, were ignored in the confrontational atmosphere at Vienna, much to Africa's disadvantage.[18] Reliance on massive new injections of aid, whether coerced by confrontation or coaxed by persuasion, is both unrealistic for Africa and a deviation from pressing problems.

All of Africa, including SA, imports mostly proven, second or third generation technologies. The one important exception is pharmaceuticals used for human drug trials in Africa against leprosy, bilharzia, malaria, onchocereiasis and other diseases which need in situ research. Even patented proven technologies are usually available from several suppliers and there are few monopolies on technical exports to Africa. The problem for most African countries is to acquire the indigenous science and technology capability to be able to screen intelligently and select from several technological choices. The 'unpackaging' or 'unbundling' of technologies by a prospective user depends on the ability to evaluate from what is internationally available.[19] This, in turn, means a science and technical information service, library, documents centre, technicians and computer networks. Most African countries neither need, nor can afford, this kind of technology transfer screening; nor will they be helped by UN Centres thousands of miles away, or by Codes of Conduct. Nigeria, Algeria and a few other countries can establish national technology transfer processes; the others need to organize critical mass groups of national R & D personnel in key fields such as computers, who can advise governments on what imports are available and desirable. The technology transfer problem is simply less relevant to Africa than it is to Asia and Latin America. It has been oversold by certain UN agencies.

Research Priorities

Avoiding tempting pseudo-priorities like technology transfer, Africa needs to get down to the real job. The priorities advocated here are:

1. Introduction of science and technology into primary education.
2. Generation of demand for R & D by foreign and locally-owned private sectors, parastatal enterprises and community groups.
3. Facilitating at the national level, research environments through publications, meetings, prizes for research, and similar activities.
4. Concentration on new R & D institutions such as technical schools, and reorganization of government research units with regional disciplinary centres of excellence for graduate university work.

Related to these priorities are the scientific and technical opportunities. While breakthroughs may occur, progress will mostly come from sustained, incremental, long-run R & D. The most promising fields are:

1. Tropical agriculture where substantial increases in yield and storage should be possible for major foodstuffs.
2. Animal husbandry where breeding and prophylactics should increase animal protein yields and reduce disease.
3. Resource surveys through remote sensing, geology, offshore drilling and cartography to identify the extent of known resources.
4. Low-cost, small-scale energy from bio-mass conversion, photo-voltaic cells or other devices, and major hydroelectric and river basin control projects from hydrology and civil engineering.
5. Mass public health campaigns to eradicate measles, poliomyelitis, yaws, and possible campaigns for bilharzia, malaria, leprosy and onchocerciasis.

Science and technology mobilized are indispensable to African development. The promise is as great as the past has been frustrating.

Research And The African Crisis

Drought, famine, rapid population increase, declining agricultural output, and unpayable external debts are the elements of an African crisis in the 1980s that is resulting in a new evaluation of the role of science and technology. Most Africans remain dependent for subsistence and survival on the natural forces of rain, wind, climate, and soil over which they have little or no control. The mobilization of science and technology to control the environment is not occuring. The incidence of infant mortality, low life expectancy and susceptibility to contagious diseases is the most vivid proof that the human condition in Africa is degrading for many. The World Bank estimated in 1986 that one-fifth of the 730 million malnourished people in the world live in Sub-Saharan Africa. Population pressures on stressed environments are producing deforestation and threatening future famines.

It is obvious that science and technology has a role to play in coping with the African crisis but it is not clear how to establish and sustain that role. One emphasis is on acclerating basic and applied research programs outside Africa and then transferring the results. This approach is being used for both tropical diseases and tropical agriculture where basic research requires instrumentation and concentrations of researchers not available in Africa. Another renewed emphasis is on strengthening African research institutions at the regional and national levels with injections of funding, expatriate staff, and training (Table 5.3).

The concentrated global research on six major tropical diseases is beginning to show promise. There is the long-term prospect of a genetically engineered vaccine against malaria. Drug trials are in progress for a microfilaricide that attacks the worms that cause river blindness (onchocerciasis). Research continues on a vaccine for schistosomiasis and on improved vaccines for measles, diptheria, tetanus, whooping cough, polio, and tuberculosis-all major killers in Africa.[20]

There is evidence from the eleven year experience of the Onchocerciasis Control Program (OCP) in the multi-national Volta River Valley of West Africa that applied research works. John Walsh writes that "the OCP is a rare example of the effective use of science and technology to combat a serious parasitic disease in Africa. In its first 11 years, the program has made major progress toward its primary goal of protecting people in seven West African countries against Oncho. It has also made it possible for numbers of people to move into fertile land in river valleys abandoned because of the disease."[21] The program remains foreign-funded and partly managed and has yet to be turned over to the African governments.

The possibility then is for important breakthroughs in research on tropical diseases whose dissemination and adaptation will require African R & D capabilities. Although there are medical schools conducting research in Senegal, Nigeria, the Ivory Coast, and Kenya there is little pharmacological research. The recent experience with AIDS disease has underlined the dependence of much of Africa on outside basic and applied research.[22] In order to benefit from new and improved vaccines generated outside Africa it will be necessary to improve in situ drug-testing and related capabilities.

Attempts to acclerate research on tropical crops outside Africa and on livestock have run into severe transferability problems. G. Edward Schuh, director of the agriculture department at the World Bank, reports that "there are parts of Africa where we simply don't know how to sustain yields unless we keep the land idle for awhile and let it rest. We simply don't know how to do it if we crop on a continuous basis, even with fertilizers imported from the West. We don't understand the soils of Africa very well or the complex of interactions among soil, rainfall, climate, and temperature."[23] There is no substitute for or shortcut to sustained in situ research on African agriculture and livestock.

Much of Africa has not experienced the Green Revolution and is unlikely to do so unless its agricultural research and extension capabilities are vastly improved. Where high-yielding breeds of maize have been locally adapted and then widely disseminated in Kenya, Malawi, and Zimbabwe it has been the result of decades of local research. There is promising research on millet and sorghum in the Sudan, and on disease-resistant cassava in Nigeria but these upgraded major food crops are not yet ready for dissemination. Imported wheat and rice have not done well in Africa and research on local varieties lags. Dr. Noel Vietmeyer has identified a number of lesser-known plants of potential use in Africa but where are the scientists to research them?[24]

African governments have been spending proportionately half as much on agricultural research as governments elsewhere.[25] Where economic growth over 25 years has stayed ahead of population increase in Africa without petroleum exports it has been in countries such as Ivory Coast, Malawi, Kenya, Zimbabwe, and the Cameroon that have invested in agricultural research and researchers. Agricultural economists Carl Eicher and Doyle Baker of Michigan State published in 1982 A Critical Survey: Research on Agricultural Development in Sub-Saharan Africa.[26] It documents what is and is not known and the research resources available at that time.

Will African governments and external donors respond to the crisis with a sustained commitment to the establishment of indigenous research capabilities, especially in agriculture, livestock, and public health? This entails a profound re-examination of non-working institutions-universities, government ministries, parastatal organizations. Another round of attempts to create national science and technology policies as in the 1970s is not needed. Institutional innovation as in the OCP program in West Africa or the Eastern Africa Desert Locust Control Organization is more appropriate. The embryonic professional associations of engineers and scientists may have an important role to play at the national level in catalyzing such innovations. External donors need to understand that human resources matter more than bricks and mortar and that the primary task is to facilitate an environment in which Africans can do sustained research.

It is ironic that 25 years after independence institution-building for R & D remains a priority.[27] Technology transfer cannot work in Africa without _in situ_ research. The challenges of tropical diseases, agriculture and livestock can only be met by in-house capable researchers at the national level even if in smaller countries this means only one or two research teams. The scientific breakthroughs may occur outside Africa but they will not be disseminated unless and until Africa acquires its own capabilities.

NOTES

1. World Bank, _Acclerated Development in Sub-Saharan Africa_, Washington, DC, 1981, pp. 69-76; John W. Forje, _Bibliography on Science and Technology for Development in Africa South of the Sahara_, Research Policy Institute, University of Lund, Lund, Sweden, 1982.

2. Aaron Segal, _U.S.-South Africa Relations: The Technological Factor_ (Bloomington: African Studies Program, Indiana University, 1984).

3. UNESCO, _Survey on the Scientific and Technological Potential of the Countries of Africa_, UNESCO Field Science Office for Africa, Nairobi, 1970.

4. Aaron Segal, _U.S.-South Africa Relations_, op. cit. John A. Marcum, _Education, Race and Social Change in South Africa_ (Berkeley: University of California, 1982).

5. UNESCO, _Mineral Map of Africa_, 1976; William Hance,

The Geography of Africa (New York: Columbia University Press, 1970); Thomas R. DeGregori, Technology and the Economic Development of the Tropical African Frontier (Cleveland: Case Western Reserve, 1969).

6. UNESCO, Science and Technology in African Development, Science Policy Studies and Documents No. 35, CASTAFRICA, Dakar, 1974, p. 83.

7. Ralph A. Austen, Daniel Headrick, "The Role of Technology in the African Past", paper presented at the annual meeting of the African Studies Association, Washington, DC, 1982.

8. Science Education International Clearinghouse, University of Maryland. The Clearinghouse maintains a database of science education projects.

9. World Bank, Accelerated Development in Sub-Saharan Africa, Washington, DC, 1981, pp. 37-44 on the public sector.

10. The Algerian corporations contracted during the 1970s with US educational consulting firms to establish and staff the two institutes while training Algerian graduate students abroad. The language of instruction in the institutes is English.

11. World Bank, Agricultural Research, Sector Policy Paper, Washington, DC, June 1981. The World Bank established in 1981 the International Service for National Agricultural Research (ISNAR) to assist technically.

12. UNESCO, Science and Technology in African Development, Science Policy Studies and Documents No. 35, CASTAFRICA, Dakar, 1974, p. 21.

13. Clement Henry Moore, Images of Development, Egyptian Engineers in Search of Industry (Cambridge, Mass.: MIT Press, 1980).

14. Bulletin of the Regional Office of Science and Technology for Africa, UNESCO, Nairobi, 1980. The American Association for the Advancement of Science (AAAS) co-sponsored a December 1984 meeting in Abidjan; The Role of Scientific and Engineering Societies in Development (Washington, DC: AAAS, 1985). It has also published a Directory of Scientific and Engineering Societies in Sub-Saharan Africa, December 1985, and is exploring assisting the launching of an African Journal of Science and Technology.

15. The Canadian International Development Research Center supported the Mauritius library project. Personal communication from Cyril Treister.

16. C. Anthony, I. Gruhn, U.S. Policy on Science for Development in Africa, UCLA African Studies Paper 19, Los

Angeles, 1980.

17. Isebill Gruhn, Regionalism Considered: The ECA (Boulder: Westview, 1979).

18. World Health Organization, 2nd Annual Report, UNDP/ World Bank/WHO Special Programme, Action Against Tropical Diseases, 1978. The WHO special programme has received funding from an international consortium and is supporting important research on malaria and other diseases.

19. Constantine Vaitsos, Inter-Country Income Distribution and Transnational Enterprises (Oxford: Clarendon Press, 1974).

20. New York Times, May 27, 1986 and December 20, 1985.

21. John Walsh, "River Blindness" A Gamble Pays Off," Science, Vol. 232, 23 May 1986, pp. 922-925.

22. Colin Norman, "Politics and Science Clash on African AIDS," Science, Vol. 230, 6 December 1985, pp. 1140-1142.

23. New York Times, August 20, 1985.

24. Noel Vietmeyer, "Lesser-Known Plants of Potential Use in Agriculture or Forestry," Science, Vol. 232, 13 June 1986, pp. 1379-1384.

25. World Bank, Acclerated Development in Sub-Saharan Africa, Washington, DC, 1981, p. 69.

26. C. K. Eicher, D. C. Baker, Research on Agricultural Development in Sub-Saharan Africa: A Critical Survey, Michigan State University, Department of Agriculture, International Development Paper No. 1, 1982, East Lansing.

27. Isebill V. Gruhn, "Towards Science and Technology Independence?," Journal of Modern African Studies, March 1984, Vol. 22, pp. 1-18.

6

East Asia: Pathways to Success

Wenlee Ting

The recent economic and industrial advances in several East Asian countries have generated a substantial amount of interest and discussion on the driving forces behind the impressive progress. Why has the East Asia region been able to chalk up such phenomenal gains while the rest of the developing world, especially in Latin America and a major part of Africa are either languishing or backsliding economically? Indications appear to point to a combination of factors ranging from economic, political to socio-cultural forces. A major line of discussion centers on cultural dimensions, especially the Confucian traditions of the more successful countries in East Asia.[1] Other observers tend to see socio-economic forces as predominant.

The question is crucial for policy implications especially in terms of the issue of the transferability of the East Asian experience to other regions. For example it would have particular relevance for the recent U.S.-backed Caribbean Basin Initiative. If culture is the major determinant of the East Asian success, then there would be scant hope of borrowing anything useful from the Asian lessons. Evidence seems to suggest otherwise.[2] An in-depth discussion of the interactions between culture on the one hand and technology and industrial development on the other is outside the scope of this chapter. The process of interactions between culture and technology is a complex one and cultural dimensions do appear to exert some form of moderating influence on the transfer, implantation and indigenous development of technology. However, technology, especially market-oriented technology, seems to have a life of its own in interacting with and impacting on successful industrial development.

This chapter will therefore focus on technology as a

major force in its own right in influencing industrialization for the East Asian countries. It will offer some perspectives on the East Asian experience from which practical, non-cultural and hence transferable implications can be drawn for other socio-cultural systems with the same industrial aspirations. Even if some form of technology development formula can be identified in the East Asian experience, applying the formula to other aspiring countries may not be as straightforward as policy makers would like but at the very least it can provide some minimal guidelines for technology policies in the non-Asian region. East Asia is defined here as that region which includes China, the Koreas, Japan, Taiwan, Hong Kong, Thailand, Malaysia, Singapore, the Philippines and Indonesia. Together, this region constitutes almost 2 billion in population and accounts for approximately 38% of the total world's GNP.

Analytical Approaches

The focus is on both macro and micro level technology dimensions in order to analyze environmental as well as firm-level variables. It is important to draw a sharp distinction between macro and micro level technology dimensions because of the tendency to subsume one for the other. On the macro level, the science and technology policies of the respective governments act as parameters within which industrial technology development has to take place. In most cases the public sector science and technology policies form a framework almost independent of the activities of private sector science and technology. Usually there is little or no productive interaction between the two despite deliberate attempts to promote such. In most non-market oriented countries like mainland China, and the Soviet Union, for instance, their rather impressive basic science and technology capability does not translate into marketable and industrially competitive technologies. These non-market technological systems have produced their share of impressive contributions to basic scientific research but these are also precisely the countries that are now hungering for just the type of market-oriented technologies that seem to spew forth effortlessly from such dynamic technological systems as those of the U.S. and Japan. In addition, India, prior to the recent changes initiated by the new Rajiv Gandhi regime, serves to illustrate the problem of a centrally dictated technology environment. For many years, India had instituted a tight system of controlling, monitoring and allocating incoming foreign technologies. As a result of such centrally mandated technology policies, the country effectively distanced itself from the system of free international flows of technology. The departure of IBM and Coca Cola in the 1970s are cases in point.

As Figure 6.1 shows, if the final technological outputs of the system are to be better performing and technologically superior products, the national technological system should have an efficient delivery system that will transmit the knowledge-embodied knowhow to product-embodied innovation. In non market-oriented systems, the absence of any substantial linkage between the basic science and technology systems and the technology output results essentially from the lack of any delivery or feedback systems intervening between them. On the other hand, the private sector firm-level technology and product development systems are directly linked to the system output of product innovations. The markets send the signals as well as act as the final users of the technological innovation for the private sector innovators.

In market-oriented systems, the strategy is to emphasize the set of relationships represented by A, i.e. the market-induced technology and innovation process. Most of the momentum of the technology policies under such a framework is to promote the establishment of a vital and dynamic factors market, especially that of technology inputs and the stimulation of a vibrant end-users of consumer market for the system outputs. In a non-market system, on the other hand, the strategy is to artificially create a channel from the public-generated technology to final system output, ie. the set of relationship represented by B in Figure 6.1. Although occasional successes are possible, the system is obviously immensely costly and precariously supported by an inefficient system of government subsidies. The following sections examine how the East Asian Newly Industrializing Countries (NICs) successfully employ the market relationships between technology development and product innovations in their industrialization strategies.

The NIC Technology Model

The industrial success of the East Asian newly industrializaing countries (NICs), can largely be attributed to the adoption of technologies policies that are essentially type A relationships and processes as shown in

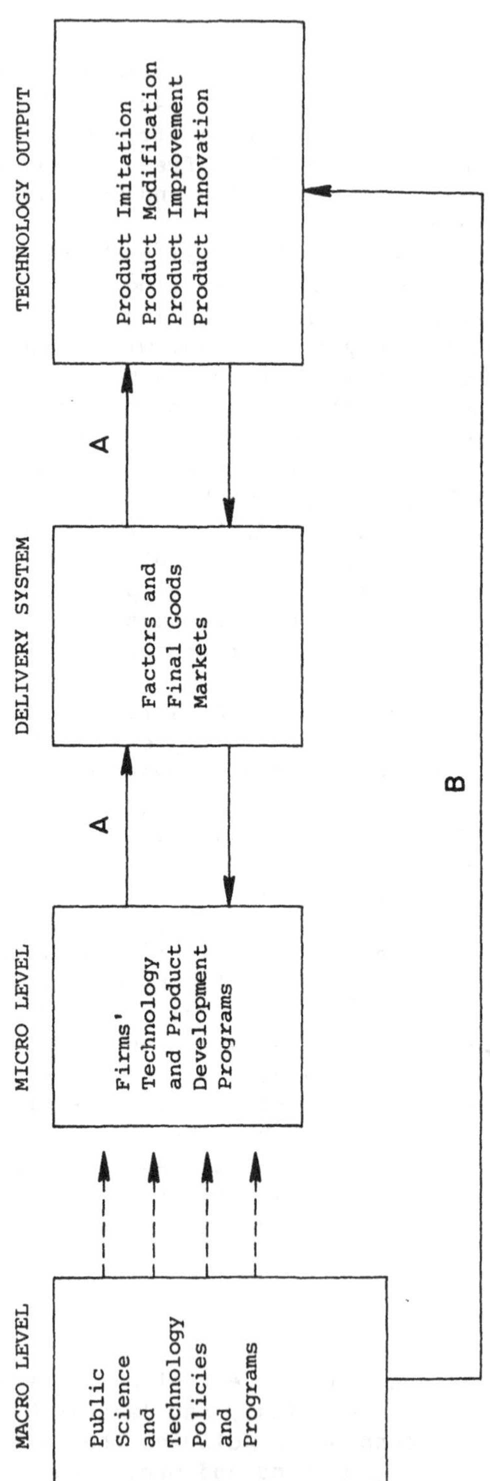

Figure 6.1 Technology Input and Output Development Framework

Figure 6.1. The technology policies pursued by the NICs are based on a strategy of pro-actively managing and stimulating evolving private market forces. Instead of stifling technological development by setting up top heavy science and technology bureaucracies and imposing expropriatory actions on market-based technology transmissions and diffusions, the NICs fostered an initial policy of attracting private foreign direct investment and facilitating the implantation of the attendant technologies. Such a policy of promoting market-based technology directly stimulated the development of a self-sustaining indigenous market for technological inputs like technical personnel and knowledge-intensive production knowhow. The development of the factor market for technological inputs in the NICs followed a multi-staged process consisting of: 1) Labor-intensive export-oriented strategy, 2) Indigenous technology development strategy, and 3) Full-fledged and self-sustained technology development and innovation.

Initial Technology Policies

Initially, in the early to late 1970s, the key aspect of the strategy of pro-actively stimulating and managing the technological inputs market was a very successful attempt at attracting export-oriented and labor-intensive technologies that would secure for the country a competitively advantageous position in the global markets. Elaborate entry management systems were set up by the East Asian NICs to attract the technology-rich multinationals (MNCs) direct investments and then to facilitate their start-up and operations. Apart from the technology transfers accomplished through direct investments by the multinationals, the international purchasing activities of major mass-marketing chains such as that of Sears and K-Mart for the U.S. markets were a fertile source of technological knowhow for the East Asian NICs. This was one of the major conduits for the transmission of consumption standards from the advanced markets to the NICs producers.

In fact, as later sections will show, the push toward indigenous technology development was facilitated more by the major marketers than by the direct investment of the manufacturing subsidiaries of the MNCs. The transmissions of product technology by foreign buyers are less formal than the more or less intact transfer of company-mandated production systems of the direct investors. As such, with a lesser amount of technological "hand holding," the local firms have a greater opportunity to develop their own production techniques and therefore lessen the dependence on foreign production technology. This was then the initial labor-intensive phase of the NICs' industrialization and technology development, a phase characterized by a predominant reliance on the multinationals and the mass marketers as sources of technology transmissions.

Indigenous Technology Development

The second stage in the development of the technological inputs market is the promotion of a self-sustaining and viable indigenous technology sector. This second stage strategy was inevitable in the face of the ever-changing, dynamic and constantly shifting global competitive advantages. As other industrially emerging nations like China, Indonesia and Sri Lanka are beginning to position themselves along the lower end of the labor-intensive competitive advantage spectrum, the NICs have really no choice but to move several rungs higher up the technological ladder. Again the two options available to the NICs were either to continue to be foreign-dependent or to proceed with indigenous development. The former would call for a strategy of continued reliance on multinationals-transmitted technologies. The more logical and perhaps rational option in the face of dynamic competitive advantages would be the pursuit of indigenous technology development. Instead of relying predominantly on what are essentially MNC-based and mostly labor-intensive assembly technologies, the NICs decided to proceed several steps further in promoting the integration and indigenous diffusion of the transferred technology and subsequently the development of indigenous technology and product innovations.

Therefore NICs like Taiwan, South Korea, Hong Kong and to a lesser extent Singapore decided to initiate and later succeeded in the early 1980s in developing a rather self-sustaining and efficient indigenous technological inputs market which provides the foundation for more advanced technological development. This strategy of deliberate indigenous technology development is basically what separates the East Asian countries that have taken off industrially and those that have not.

A case in point is the strategy adopted by Malaysia and Singapore as passive recipients of labor-intensive electronics assembly operations of major multinationals. Both countries strive aggressively to become captive assembly arms of major American electronics companies which were naturally attracted to the well-administered and infrastructurally-efficient countries. Malaysia, for instance, has the dubious distinction of being the largest exporter of off-shore semiconductors in the world. In a typical international semi-conductors production system, the more demanding design and fabrication of the chips are performed in the U.S. while the highly labor-intensive assembly of the sub-components are then assigned to cheap labor-cost countries like Malaysia and the Philippines with higher skilled testing being done in Singapore and Taiwan. The semiconductor devices are then shipped back to the U.S. for finishing and final shipments. Virtually all the chip outputs imported into the U.S. in the early 1980s were sourced from wholly-owned U.S. manufacturing subsidiaries in East Asia while negligible amounts are sourced from independent local suppliers in these countries. Such operations generated a great deal of unskilled and semi-skilled employment but little or no technology transfers in the form of self-sustained acquisition of technical skills by the local entities.

The more advanced design and fabrication work is still done primarily at home. The recent plant closings by several U.S. electronics makers in Malaysia and the resistance of multinationals in Singapore to upgrade the technological contents of their operations have caused serious setbacks in the industrialization plans of these two countries. Since Malaysia had not made substantial efforts to foster indigenous technology development, its strategic options in the face of the pull-out by the multinationals were therefore severely limited. Singapore's over reliance on MNC-based technologies and hence being subject to the dictates of the foreign multinationals is one of the major reasons behind its recent economic difficulties. The country is now currently attempting to rectify the situation by making strenuous efforts to diversify out of its MNC conundrum.

Full-Fledged Technology Development And Innovation

The NICs are currently striving to move out of the second and into the third stage of self-sustained technology advancement. This phase which is still in its earliest infancy has nevertheless been marked by several relatively notable successes, especially in consumer electronics and automobiles in the case of Taiwan, South Korean and Hong

Kong firms. This stage has also begun to witness the rather forceful attempts at internationalization by several more aggressive NIC firms particularly from South Korea. A notable case in point is the recent small car export efforts of the South Korean Hyundai conglomerate to enter the U.S. and Canadian markets. In Taiwan, the abundance of high technology enterpreneurial firms enables the country to position itself on the forefront of consumer electronics manufacturing. For instance many brand name U.S. personal computers and computer parts sold in the American market are Taiwan made. In addition to being the original equipment manufacturers (OEMs) for major firms like ITT and IBM, Taiwan electronics companies are venturing into the computer peripherals markets in the U.S. with their own in-house brands and some have established their own market delivery systems by teaming up with independent computer retailers in the U.S. At the same time a few engineering-oriented computer firms in Hong Kong are coming out with IBM-compatible computers based on circuit-boards and disk drives that are designed in-house.

The Product Technology Innovation Continuum

The following sections provide a more detailed perspective of the stages of technology and product development of the East Asian NICs and other industrially aspiring countries of the region. A dynamic framework for the technology development and product innovation process is appropriate in analyzing the technology development trends in East Asia. Such a dynamic framework is provided by the product technology innovation continuum in Figure 6.2.

Because of the rapid changes in the East Asian industrializing economies, a static look at the technology dimensions of the different stages of development would obscure many of the still evolving trends in the NICs' development process. The product technology innovation continuum or PTIC enables the analyst to focus on the developing trends instead of merely static and restricted snapshot views of the development activities. The PTIC is therefore appropriate for examining the technological trends by allowing the incorporation of new realities and the projecting of future scenarios. By focusing on the past, present and future technology strategies of the NIC firms, the PTIC framework highlights the contrasting roles of both MNC-based and indigenous technology development.

The framework captures the technological dynamics

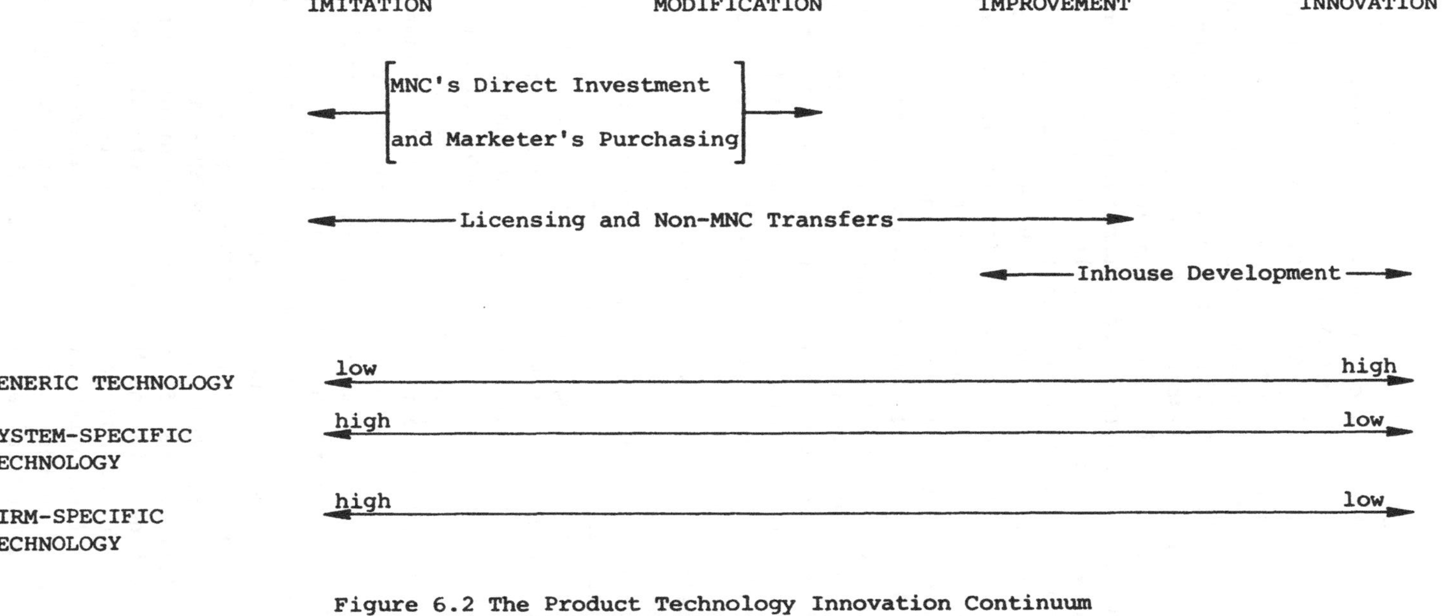

Figure 6.2 The Product Technology Innovation Continuum

facing the East Asian NICs through identifying 4 major interconnected stages: product imitation, product modification, product improvement and product innovation as represented in Figure 6.2. As the figure shows, at the firm level, a company must proceed through 4 stages of development. Progress through the stages requires the utilization of different forms of technology, ranging from firm-specific and system-specific technologies to generic technology. Firm-specific and system-specific technologies are the most clearly defined and established. The entire system is known and pre-determined in terms of set company procedures and techniques. In the case of generic technology and knowhow, on the other hand, only the general parameters of the theoretical knowledge are defined while actual operating systems remain unspecified. As the PTIC indicates, the technology development made during the product imitation and modification stages is mainly dependent on the transmission of firm-specific technology originating from the investing multinational firms. Under these two stages, the key consideration is exact duplication of existing products by labor-intensive methods for the MNCs to be re-exported back to the companies' home markets, say, in the U.S. or Europe.

The strategy here calls for combining the multinationals' need for low labor costs and the host NIC's need for export development. From the standpoint of factor market development, there is very little or no development in the indigenous technological input markets. Due to the high firm-specific and system-specific technologies of the MNC technology, only a small amount of low-end labor-intensive skills, such as in routine assembly of components, are imparted to the local production workers. A more important contribution to the technological development of the host country is the exposure of the workers to a regimen of industrial discipline rather than any substantial imparting of skills through learning-by-doing.

The transmission of higher and more sophisticated skills that could lead to the emergence of a generic and independent technology system is stymied by the firm-specific and disaggregated nature of the MNC-based technology. Many of the off-shore manufacturing facilities of the multinationals in the NICs are essentially captive production arms of the multinational companies with a product mandate often oriented to the home market. However, a major exception to the lack of indigenous development during the technology imitation stage is when the multinationals consciously and deliberately attempt to develop sources of local inputs. Professor Chi Schive for instance has documented a substantial amount of skill transfers by multinationals in developing local vendors for critical sewing machine components in Taiwan in the early seventies.[4] The same pattern of skill transfers appears to have taken place in Taiwan's electronics industry. It is perhaps from this small beginning that the foundation for Taiwan's electronics and sewing machinery industries was laid. The profusion of small entrepreneurial firms that sprung up to supply the foreign investors and the buyers of the mass marketers succeeded to a great extent in providing the springboard for the not unimpressive stage of technological achievement evident in Taiwan today. Thus whenever local supplier development exists, many small technical-oriented entrepreneurial firms are created locally. This is a necessary pre-condition for subsequent progress into the more advanced stages of technology development and product innovation.

The more advanced stages of product improvement and innovation in the PTIC require different technological processes in the development of the local factor market for technological inputs. Under the product improvement and innovation stages a more demanding environment is necessary in terms of developing more sophisticated and generic technological skills. In these two stages, the NIC firms' strategy is to go beyond the passive reception of MNC-transmitted production knowhow and instead develop a set of independent and indigenous product development skills. As noted earlier, such a drastic departure from MNC-sponsored technology is necessary to keep the firms competitive in their own right in the ever more intense world market place. It also releases them from the product mandates imposed by the multinationals. The shift from MNC-based technology to the acquisition of generic technology skills is contingent upon the establishment by the host country of a cost-efficient and market-directed technological structure that would span the entire spectrum of technological knowhow from low-level labor-intensive skills to engineering and design skills.

The successful establishment of such a market-oriented technological structure to promote the development of non-MNC, generic and indigenous technology is dependent upon meeting several necessary pre-conditions: (1) the creation of an easily available, well-developed and relatively inexpensive supply of local engineering and design skills, (2) the acquisition of an appropriate type of technology, (3) the setting up of a resource-appropriate production system and (4) the initiation of positive government efforts to support the development of indigenous technology via the pro-active management of market forces.

Creation Of A Supply Of Technical Personnel

For the first pre-condition, that of creating a large and available supply of engineering and technical personnel, many of the NICs have initiated well-implemented secondary and tertiary technical education programs that not only supply lower and middle level technical expertise but also higher design and engineering skills as well. At the tertiary level, the establishment of advanced training institutes, funded partly by foreign governments and by major multinational corporations is also contributing to the supply of mid-level manpower. Both South Korea and Taiwan have established advanced scientific research institutes to help spearhead the technology development efforts. Another major development has been the "reverse brain-drain" that is gradually gaining momentum as the East Asian NICs begin to establish technological infrastructures capable of absorbing their highly-trained Western-educated graduates. The recent rapid pace of industrialization and the opportunities it generated has attracted a return flow of Western-trained engineers, especially to Taiwan and South Korea. In Taiwan, for instance, the returning graduates, many with masters and doctorate degrees in electrical engineering, have been instrumental in stimulating the establishment of many high technology entrepreneurial firms.

The secret behind the rapid and successful technology development in the NICs, particularly Taiwan and Korea, is not the mere quantity of the formally trained engineers and technicians that these countries graduate every year, but rather the effective integration of these personnel into the endogenous technological production systems. While government-sponsored formal education and vocational programs contributed substantially to the pool of qualified manpower in the NICs, it is the private entrepreneurs and firms which mobilized the productive potential of the pool of technical talents essentially through on-the-job training. Except in Singapore which has rather elaborate formal joint government-private sector training programs. Taiwan, South Korea and Hong Kong primarily pursue an informal learn-on-the-job training strategy. Formal and highly organized training programs are mainly tailored for standardized mass-production systems. Since most of the

NICs have highly organic, flexible and "unstructured" production systems, formal training in organized production skills may seem irrelevant to job effectiveness.

Acquisition Of Appropriate Technology

Secondly, at a critical juncture of their technological development, the East Asian NICs were further helped along by the advent of the electronics revolution in the late seventies and early eighties, a revolution that was characterized by product technologies with rapidly changing and increasingly short product cycles. The aspiring NIC manufacturers cannot have asked for a better coincidence. The electronics technologies have the advantage of being knowledge-intensive rather than capital-intensive. No large scale investment in expensive production systems was necessary to initiate manufacturing, especially in the design of end-user items like computers. The advent of such knowledge-intensive technology enabled many technically capable but cash-short entrepreneurial firms in Taiwan and Hong Kong to venture into commencing business. Knowledge-intensive technologies requiring very little firm-specific or system-specific technological inputs are most conducive to the development of indigenous technology from scratch. All it requires is a set of generic technological skills capable of being applied to product development experimentations. In addition, the electronics industry's short and rapid product cycles, although a source of risk for the well-established capital-intensive firm, came as a boon for new entrepreneurial firms because the constant shifts in technologies and breaks in rapidly moving product cycles provide many opportunities for the newcomers that would not have been otherwise possible in more rigidly established industries.

The widespread and rapid diffusion of consumer electronics in East Asia appears to have been facilitated by just such a set of characteristics. The combination of the engineering and design-intensive nature of the electronics technologies and the limited need to set up expensive capital-intensive production systems basically accounted for its widespread and successful implantation in NICs like Taiwan, South Korea and Hong Kong. The electronics revolution was also propelled along by the almost exponential decrease in the price of the ubiquitous semi-conductor memory chips; with the price of the 64k chip, for example, falling from $50 in 1980 to less than a dollar in

1985. The above factors together with the high degree of adaptability and flexibility of the NIC firms production capabilities all conspired to produce one of the most rapid and dynamic developments in the experience of technological diffusion.

Setting Up Endogenous Production Systems

The third major pre-condition for the successful advance up the PTIC ladder is the setting up of an endogenous and cost-efficient production function or system that will optimize the resource and organizational infrastructure of the local economy. Many of the newly emerging technology entrepreneurial firms lack expertise when it comes to establishing manufacturing systems. Unlike the earlier era of MNC-based transfers, the emerging mode of indigenous technology development has to proceed without any firm-specific and system-specific technological guidance. The lack of such technological guidance has actually came as a disguised blessing. The NIC firms, especially the smaller entrepreneurial concerns, have to set up from scratch rudimentary production systems through trial and error. The production system of these newly emerging NIC firms were no longer the extensions of MNC-mandated systems nor the production arms of mass-marketers from the advanced markets. Instead they are increasingly locally conceived and designed systems that were derived from more or less generic technological knowhow.

Such knowledge-intensive but capital equipment poor production systems were easily deceptive to uninitiated observers. Many of the developing NIC firms were perceived to be operating at low technological and capital-intensive levels, especially when measured in terms of classical capital-labor ratios. The use of a conventional capital-labor intensity measure of the technological level missed precisely the true orientation of the production systems of the East Asian technology firms. In studies of the production functions of the then developing NICs firms, many of these firms are showing signs of departing from the generally perceived stereotype of low capital-labor make-up in their production function.

The seemingly low technological level of the NIC manufacturing systems yet combined with their rather impressive level of product quality and technology is one of the many paradoxes in the technology development experience of the East Asian firms. NIC electronics, fashions, toys and household appliances marketed in the U.S. and other Western markets are surprisingly well accepted by more sophisticated consumers. The quality and the technological level of the NIC producers are better reflected in pirated computers and counterfeit designer watches and other look-alikes. These ironically provide a better measure of the emerging technological and engineering capabilities of the NIC firms than any capital-labor ratios. As noted, one of the keys to an appreciation of the impressive technological achievements of the NICs is an understanding of their production systems. What then are the kind of production systems that can actually account for the discrepancies between perceived technological levels and the output quality of East Asian NICs products?

Managed Improvisation And X-Efficiency

The reality behind the apparent paradox is a unique approach to the organization of production among NIC firms especially in Hong Kong and Taiwan. Essentially the approach calls for the utilization of a production function in which capital, labor and technological inputs like design and other manufacturing skills operate under a system of flexible, interchangeable and interactive uses. In addition, the manufacturing functions are not centralized under any single and rigid firm entity and are instead dispersed amony many independent but interactive production units. This interactive, flexible and "organic" form of organization of production units can best be conceived of as "managed improvisation."[5] Under a production system characterized by managed improvisation, production inputs, ranging from capital equipment to multi-skilled workers and technicians, are actively interchangeable in most stages of the production process. The production possibility curve is therefore one characterized by a variable elasticity of substitution. In the multiple uses of capital, for example, the same piece of equipment is applicable to different functions and operations at the various stages of the manufacturing process. The production and assembly lines of many NIC firms are therefore less continuous and more intermittent in workflow than that of comparable U.S. and Japanese systems. Even with automated systems, skilled manual sequences are frequently interposed between otherwise automated production flows. The incorporation of a high degree of people-embodied skills suggests that even without the formal guidance of a system-specific technology transmission, the NIC firms are quite capable of setting up production systems that need not depend too heavily on the specialized machinery of conventional mass-production systems. For the same level of productivity and product quality, NIC manufacturing systems require a lower level of capital in the form of expensive and specialized equipment and processes. At the same time, the system of managed improvisation based on a multi-skilled and flexible workforce and production units can contribute to a high degree of flexibility and adaptability in the NIC production systems.

The process of managed improvisation which underlies the production efficiency of NIC firms can best be explained by a concept which Leibenstein called "X-efficiency." He suggested that production functions are seldom completely pre-determined or known. Thus for a given level of inputs, additional output gains can be realized without any increase in capital or labor inputs. Or as he states, "inputs may have a fixed specification that yields a variable performance or they may be of a variable specification and yield a variable performance."[6] In other words, the production function has sufficient built-in variability to allow for potential improvement in efficiency without additional investment in capital or people. This potential output gain is termed the X-efficiency. X-efficiency may arise from various slacks in the production transformation process, such as the deficient use or underutilization of production inputs. Thus sub-optimal uses of capital and human resources could lower productivity or quality in the system output. The use of managed improvisation is to eliminate or reduce existing slacks and sub-optimization in the production process through more intensive use of capital equipment and through substituting human-embodied skills to enhance the given level of capital usage.

A recent study by Redding and Tam of the University of Hong Kong appears to confirm this flexible pattern in the production system of small entrepreneurial Hong Kong firms.[7] The sample firms were found to have achieved a certain level of X-efficiency through the setting up of flexible production systems based on managed improvisation and by avoiding a more structured and rigid production system. The production system consists of a collection of interacting and yet independently owned firms organized according to trade functions or task specialization. The system is composed of essentially two main levels: the merchandisers who provide the link to the foreign trade buyers, and the mass network of production units with their supporting layers of suppliers and vendors. The entire organic unit is linked by affective (kinships and connections) rather than contractual relationships. The merchandisers which are actually import-export entities specialize in the function of acquiring orders and then channeling the jobs through a connected network of individual firms that perform the actual production work. This modular type of production organization consisting of individual firms that interact in a collective system permeates Hong Kong and Taiwan. In fact, the modular production systems in these two NICs are actually an integral part of an international production system with front-end units represented by the mass marketers of the U.S. and other Western markets while the manufacturing supports are provided by the vast network of small producers in Hong Kong and Taiwan.

As noted earlier, the existence of such a system greatly facilitated the cost-efficient development of indigenous technology by NIC firms. First, a mass network of independent and individually-owned firms avoids the burden of overhead costs and capital expenditures involved in setting up an in-house production system under the entity of a single firm. Secondly, the NIC producers are not subject to constraints imposed by firm-specific MNC-based transfers. The imposition of product and technical mandates by the multinationals on their suppliers tends to preempt any opportunity for the local firms to take fresh product initiatives which hence reduce their innovativeness. Local suppliers to the multinationals are operating under the dictates of an already established production system. In the modular type set-up, the individual firms are not tied into any pre-determined system- or firm-specific constraints and are therefore free to take greater production initiatives on their own accord. Thirdly, the entire system benefits from greater competition among the vast array of small manufacturers. It has been estimated that about 40,000 small companies produce 80 percent of Taiwan's exports.

However, there are also disadvantages to the modular type set-up. The major disadvantage is that the small size of the firms precludes any strategic and marketing clout in a world market place increasingly dominated by large international corporate conglomerates. A related disadvantage is the lack of any marketing identity. The small firms will probably remain submerged under layers of anonymity and may at times be captive to the front-end mass marketers.

The approach to organizing production systems in Singapore is somewhat different from that of Taiwan and Hong Kong. As Singapore has aggressively courted the major multinationals and has adopted a strategy of offering captive production arms to the foreign investors their production systems are therefore essentially off-shoots of the Anglo-American type. The major advantage of such systems is the access to and integration with the global system of MNC-based production and marketing. The major drawback is obvious. Unlike the loosely but efficiently-linked production systems in Hong Kong and Taiwan which have a greater degree of freedom in taking product initiatives, the manufacturing orientation in Singapore is predominantly tied into the product mandates of the MNC sponsors. Indigenous technology development is therefore accordingly stymied. In fact Singapore owed its rather spectacular growth more to its superior mastery of the service technologies like banking, financial services and tourism and its trading clout than any manufacturing capability. Singapore has the most sophisticated service and trade infrastructure in the East Asian region after Hong Kong.

The South Koreans' approach to technology development and the organization of production functions also differs from those of the other NICs. The Koreans strive to emulate the Japanese model in attempting to set up large scale conglomerates and mass-production systems. Vertically integrated and horizontally diversified South Korean conglomerates like Hyundai, Samsung and the Lucky/Gold Star Group are spearheading the country's indigenous technology development and attempts at breaking into the more advanced stages of product improvement and innovation along the PTIC. The Koreans are most visible in two major areas, computer electronics and automobiles. In the semiconductor field, Korean conglomerates have adopted a strategy of what has been termed "reverse technology transfers." The strategy consists of either entering into joint ventures with or buying into high technology firms such as chip design operations in mainly the U.S. By establishing a presence in especially the Silicon Valley, the Korean firms hope to attune themselves to the latest innovations and to tap into state-of-the-art technologies. For instance Hyundai has opened an electronics manufacturing facility in Santa Clara, California. Samsung has a joint venture with ITT in advanced integrated circuit (IC) devices while Lucky-Goldstar's joint venture is with AT&T in advanced telecommunication electronics. Among other examples, Zilog has licensed Gold Star in the production of the Z80 microprocessor while Advanced Micro Devices have sold 64k chip technology to the Korean company.

The South Koreans may face tremendous risks in their high-stake technology gamble. For instance, their aggressive thrust into chip making is exposing the firms to substantial risks. Chip making is highly dependent on capital-intensive and expensive sophisticated fabrication facilities. The succeeding generations of chips are characterized by increasingly shorter product cycles and rapidly falling prices. For instance, Hyundai and Samsung came out with their first output of 64k D-RAM (dynamic random access memory) just when the market was being overtaken by the 256k D-RAM in which the Japanese manufacturers dominate (80 percent of worldwide market share). The process may be repeating itself as the Japanese continue to cast a long technological shadow over their Korean counterparts in the production of the 256k chips. The 256k D-RAM is believed to have an even shorter life than the 64k chips and major Japanese makers like Hitachi, Fujitsu, NEC and Toshiba are planning to ship megabyte superchips (capable of holding a million pieces of information) in 1988 at about the same time Hyundai and other Korean makers will be coming out with their 256k chips. South Korea's headlong rush into capital-intensive and heavy-industry-based industrialization was accomplished at enormous cost. Toward the mid eighties, the economy was beset by a highly leveraged corporate sector and a huge external debt of over $40 billion, one of the highest in the world.

Changing Technological Contents of Exports

Regardless of the different approaches taken by the various East Asian NICs in the organization of their endogenous production systems and their technological development, their progress up the various stages of the PTIC appears to be reflected by the changing contents of their exports. Table 6.1 depicts a pattern which shows an increase in the export of high-technology items for NICs like Hong Kong, South Korea and Taiwan, as compared to that of other non-NIC countries such as Indonesia, Thailand and the Philippines. Using 2-digit SITC code, basic manufactures such as leather, paper and textile products were designated as "low-technology" while machines, tools, electrical and electronic goods such as television, appliances, electrical machinery and semi-conductors were designated as high-technology. From the 1966/70 period to

TABLE 6.1
Changes in Technological Composition of Exports, 1966-1981 (as percentages of total exports)

Average for:		1966-70	1971-75	1971-80	1981
Hong Kong	Low-Technology	13.3	11.8	9.4	10.7
	High-Technology	10.2	14.2	16.8	20.5
Taiwan	Low Technology	29.0	24.8	23.6	22.8
	High Technology	13.4	21.3	23.4	25.6
South Korea	Low Technology	24.9	20.1	18.6	16.7
	High Technology	6.6	17.0	14.9	15.0
Thailand	Low Technology	1.2	5.2	7.3	10.9
	High Technology	0.37	0.67	3.3	4.7
Indonesia	Low Technology	N.A	N.A	0.36	1.1
	High Technology	0.75	0.28	0.41	0.34
Philippines	Low Technology	5.3	9.4	4.8	4.6
	High Technology	N.A	6.1	1.5	2.6

Sources: United Nations International Trade Statistics Yearbook, 1983; Statistical Yearbook of the Republic of China (Taiwan), 1983, Executive Yuan, The Republic of China.

1981, for example, it can be seen that high technology exports from Taiwan, as a percentage of total exports, rose from 13.4% to 25.6%; while those of Hong Kong rose from 10.2 to 20.5 and for South Korea, from 6.6 to 15.0%. During the same period, the low-technology contents of the NICs' exports declined slightly. The pattern for the non-NIC countries shows both low-high-technology items as only a small fractional percentage of total exports. With the exception of a slight increase in low-technology exports from Thailand, manufacturing exports from the others showed no appreciable increase in either category. This trend of development appears to provide some evidence that the NICs have attained technological progress through moving up the PTIC while the non-NICs, at least as measured by export contents, have yet to take off industrially.

Government Technology Promotion Strategies

The following sections examine the role of the public sector and government efforts in helping to promote technological development in the NICs and other East Asian countries. Many observers instinctively look for grand government schemes in promoting technological development through supporting elaborate scientific research, providing direct tax and financial incentives and pouring money into education and training programs. With the possible exception of the close government-private sector collaboration in export promotion in South Korea, the whole thrust of the successful technological efforts in the NICs stems from the market forces unleashed by private enterprise. Nothing exemplifies this better than the "positive non-intervention" of the Hong Kong government. In Taiwan, the only "big" public-sector money that has to do with development went into infrastructure projects. Thus, it should be remembered that most of the government efforts are merely supplementary to and reinforcing of the market-oriented technology development carried out essentially by private enterprises in these countries.

A quick impressionistic survey of the Asian NICs and other developing countries showed varying degree of government involvement in promoting the development of science and technology for industrial development. The NICs lead the way in the government providing a guiding role for private sector technology development. In Taiwan, for instance, apart from the usual ministerial agencies and advanced research institutes conducting basic scientific studies, there is no formal national policy of fostering technological development such as in the case of mainland China with its publicly proclaimed "Four Modernizations" programs. The only sector in which the government-level policies played a rather significant role is in agricultural development. Taiwan today stand as a recognized leader in achieving impressive gains in crop cultivation technology and farm management techniques which resulted in a level of rural productivity unmatched by many other developing countries. The rapid advance in farm technology has produced a level of farm productivity which was transfered into the industrial sector and this facilitated the rapid restructuring of the economy from one which is mainly agrarian-based to one which is manufacturing-based. The same process of advancing agricultural technology appears to have spurred industrial development in South Korea. In both Taiwan and South Korea, therefore, advances in agricultural technology serve as one of the crucial pre-requisites for subsequent industrial development.

Outside of agriculture, public sector-based technological activities in the form of research institutes and government programs are however relatively less active in other non-industrial disciplines such as medical research and the physical sciences. Very little or none of Taiwan's recent gains in industrial technological capability emanated from basic research and development produced by indigenous sources. The government however plays a very active role in fostering high technology development based on adapting and developing acquired technologies. The Electronics Research and Service Organization (ERSO), for example, is aggressively spearheading the diffusion of high technology by granting patents rights to local firms for technologies it has acquired abroad or by conducting its own R & D in semi-conductors and computer software for subsequent transfer to the private sector. Its most notable recent effort in promoting high technology has been the joint set up of a completely computerized and automated circuit-board plant with Hewlett-Packard and the large Taiwan conglomerate, Formosa Plastics. ERSO's other projects ranged from the development of computer peripherals like printers and disk drives to robotics and computer-aided design (CAD) software.[8]

Other high technological efforts by the government include the setting up of science-oriented industrial zones. Following its successful pioneering efforts in the operation of free trade zones, Taiwan proceeded to create a science-based industrial park at Hsinchu with supporting quasi-government research labs and nearby science universities. The objective is to turn the zone into a high technology showcase. Since its inception, the science park has managed to attract various foreign high technology firms which have invested in such fields as fiber optics and wafer fabrication. Taiwan has plans to establish three chip making plants by the mid-eighties. In South Korea, a similar high technology promotion agency, the Korea Institute of Electronics Technology, also engages in government-sponsored electronics R & D, although the brunt of the technology development efforts, as discussed earlier, are borne by the private conglomerates.

Although many of the technologically emerging East Asian countries possess the necessary technological capability for high technological development, a supportive financial infrastructure seems to be lacking. The start-up and eventual take-off of high technology entrepreneurial firms in the U.S. as exemplified by the Silicon Valley experience was largely propelled by a well-organized and efficient venture capital financing structure. Many of today's successful high technology firms in the U.S. were initailly seeded by venture capital. Because of the financial and foreign exchange restrictions in Taiwan and South Korea, the formation of a viable equity market oriented toward venture capital financing of high technology firms does not appear a possibility in the near future. As a result, many promising ventures have to depend on more conservative funding sources such as that of commercial banks which tend to avoid risky loans of such nature.

In laissez-faire Hong Kong, there is almost an absence of government-sponsored industrial technologies, although a semi-public organization, Productivity Center of Hong Kong, plays an important role in encouraging the adoption of new products and processes by local firms. The colonial government is also gradually abandoning its previous policy of "positive non-intervention" by encouraging the setting up of electronics components operations both foreign and local. In addition, there are plans to set up several government funded R & D institutes to promote the development of, for instance, computer-aided manufacturing and computer-aided design software. In Singapore, the government role in technology development has traditionally been rather substantial. The government is directly involved in technology development through forming joint ventures with foreign partners, in addition to operating training and research institutes jointly with multinationals like IBM and Hewlett-Packard.

Among the ASEAN (Association of South-East Asian Nations) group of states, the push toward industrialization and technology development has somewhat mixed results. In the case of Malaysia, as observed earlier, efforts at promoting the development of off-shore electonics assembly for the multinationals ran into difficulties recently with the recession in the semiconductor markets. Lacking a large and self-sustaining manufacturing sector, the country has little or no alternate options in proceeding with its technology development programs. In addition, its national car program which is intended for import-substitution is temporarily impeded by sluggish demand and technical problems.

Thailand, another ASEAN member, also declared ambitious technology development plans to lift the country from its agrarian-based economy to one of high technology industrialization. Development and promotion efforts are spearheaded by the Thailand Development Research Institute, a high technology think tank and the Thailand Institute of Scientific and Technological Research (TISTR). Through partial funding by such international agencies as the World Bank and USAID (U.S. Agency for International Development), as well as from the government, TISTR plans to enter into joint ventures with private sector firms in four major areas - microwave devices, solar cells, biotechnology and computer software.

Technology development in oil-rich Indonesia follows a similar pattern in terms of attempts at restructuring a predominently agrarian economy into an industrializing one. The Philippines, much like Malaysia, had laid a rather impressive foundation in attracting major off-shore electronics assembly opeations by U.S. multinationals. However, because of its preoccupation with continuing political turmoil and its balance of payments problems, the country has relegated technology development efforts to near the bottom of its national priorities. These four ASEAN countries will probably form the next tier of newly industrializing countries in East Asia as they begin to surmount their temporary difficulties.

The People's Republic of China is probably the great unknown and perhaps the dark horse in the East Asia technology equation. (See Chapter 7) Coming from the same stock of people that have produced the NIC miracles in Taiwan, Hong Kong and Singapore, this vast nation of a billion people holds tremendous promise in the perceptions of many observers. China as a result of its recent liberalization, its experimentation with free enterprises, and its rather successful attraction of export-oriented foreign investment has already caught up with Taiwan and the other NICs in textiles, footwear, toys and plastic goods. China is replaying the NIC experience of an earlier era in pursuing similar industrial strategies and it is succeeding faster than anticipated. Replication of the NIC model of industrialization and technology development is evident in the setting up of the special economic zones (SEZs). The zones are intended to spur industrialization and technological development through the attraction of export-oriented manufacturing investment that will form the basis for subsequent stages of indigenous technology development.

Unlike the earlier experience of the NICs, however, the Chinese prefer less conventional modes of technology tie-ups. Although wholly-owned subsidiaries of foreign multinationals are permitted in the SEZs, the country still prefers more restrictive type of arrangements like contractual joint ventures and compensation trade as a means of technology transmission. Despite its very rapid progress in many product areas, serious production and management problems remain. Reports of inefficiencies, unresponsiveness of Chinese firms and workforce, bureaucratic constraints and a general lack of an industrial infrastructure may have to a large extent hampered its development efforts. The vastness of the Chinese hinterland and the problems left by decades of political and economic isolation and disruptions are formidable hurdles the Chinese have to overcome in order to smooth the path to further industrial and technological progress. The Chinese technological system is just entering into the very early stages of the PTIC.

The technological giant of Asia, Japan, is both anticipating and responding to the technological dynamics of the East Asian region with its continuing aggressive technology policies. Apart from the race with the U.S. for dominance in fifth generation computers and artificial intelligence, it is also aggressively attempting to maintain if not to enlarge its lead vis a vis its Asian neighbors.

Nothing better illustrates this continuing effort than the recent moves by Japanese regional prefectures in establishing self-contained high technology communities known as technopolises that should disperse industries away from the overcrowded Tokyo-Osaka metropolitan areas. A technopolis is an integrated living environment with social cultural amenities combined with high technology industries.[9] Nine technopolises, out of a total of 19 planned, have been established so far. The most striking example of the Japanese move in creating a network of high technology centers is the Kumamoto Technopolis on the "Silicon Island" of Kyushu. The Kumamoto Technopolis emphasizes applied and production oriented research in four major high technology areas: Automation, biotechnology, electronics and information systems. The technopolises are using a broad range of financial incentives and subsidies to attract investments from outside Japan in these high technology areas. Eventually all 19 technopolises will have an integrated community of universities, research institutes, residential and recreational facilities and industrial parks that will provide the catalyst for further technological development. Apart from a list of well-known Japanese companies, foreign firms like Dow Chemical, and Texas Instruments have established plants in the technopolises.

Summary And Policy Implications

The main patterns that can be drawn from the East Asian, especially the NICs' experience in technology development are : 1) The promotion of market-oriented technology policies both on the national as well as on the firm level, 2) the choice of a technology and production system that optimises the local technical and organizational resources and infrastructure, 3) the choice of technologies that are knowledge-intensive rather than capital intensive, 4) the adoption of a set of government policies that will stimulate the creation of a self-sustaining indigenous technological structure which can attune itself to the constantly shifting comparative advantages in the global markets. Japan and the first tier of the Asian NICs have already taken off to self-sustained industrialization. These countries are now faced with the challenge of having to constantly adjust their industrial and technology policies to attune themselves to the competitive realities of international markets. The next tier of Asian countries, the ASEAN countries, may also be poised to enter into self-sustaining industrial and technology development, if policy corrections are undertaken to set the organizations and systems in both the public and private sectors firmly on the course of technological and industrial success. Given the present state of their rather stagnant industrial development, the ASEAN nations may do well to look to their next-door neighbors, the NICs for some of the lessons of success.

NOTES

Acknowledgments: The author is very grateful for the research assistance provided by Leonard Trudo, Masood Aziz and Ning Yu.

1. Herman Kahn, World Economic Development, 1979 and Beyond (Boulder, Colorado: Westview Press, 1979).

2. Wenlee Ting, Business and Technological Dynamics in Newly Industrializing Asia (Westport, Conn.: Greenwood Press, 1985).

3. Wenless Ting, "The Product Development Process in NIC Multinationals," Columbia Journal of World Business, Spring 1982, pp. 76-81.and Ting, op. cit., pp. 78-90.

4. Chi Schive, "Technology Transfers through Direct Investment: A Case Study of Taiwan Singer," in Proceedings of the Academy of International Business, Asia-Pacific Dimensions Meeting. Honolulu, Hawaii, December 1979, p. 91.

5. Ting, op. cit., pp. 83-85 and pp. 130-134.

6. Harvey Leibenstein, "X-Efficiency versus Allocative Efficiency," in Beyond Economic Man (Cambridge, Mass.: Harvard University Press, 1980, pp. 29-47).

7. Gordon Redding and Simon Tam, "Networks and Molecular Organizations: An Exploratory View of Chinese Firms in Hong Kong," in Proceedings of the Academy of International Business, South-East Asia Regional Meeting, Hong Kong, July 1985, pp. 129-142.

8. "The King of Knockoffs Rushes to Go from Imitation to Innovation," Business Week, November 26, 1984, pp. 188-194.

9. Magoroh Maruyama, "Report on a New Technological Community: The Making of a Technopolis in an International Context," Technological Forecasting and Social Changes, Vol. 27, 1985, pp. 75-98.

7

China: The Search for Strategies

Richard P. Suttmeier

Introduction

China in the 1980's is launched on an ambitious program of economic and societal modernization. A crucial part of this program is the modernization of science and technology. China's current strategy of modernization differs in important ways from earlier strategies. Yet, a common theme running through Chinese modernization efforts since 1949 has been the central role of science and technology (S & T) in promoting economic development, national defense and cultural change. Over the years this central interest has been translated into strong government programs for a kind of state directed, "top down" approach to the development of science and technology.

One can infer from current Chinese policy discussions a set of objectives for technological development by the year 2000 which include the following elements. First, for the industrial economy in general, China hopes to upgrade the level of technology by the year 2000 to the levels of the advanced countries during the 1970's and 1980's. Second, China attaches special priority to selected areas of high technology - computers and electronics, biotechnology, materials, lasers, robotics - which its leaders believe will be the basis for wealth and power in the 21st century. In these areas, China hopes to approach the levels of the advanced countries by the year 2000. Third, there is an urgent need to develop the technologies of China's infrastructure in order to relax the severe constraints on development now felt in the areas of energy supply, transportation and communications. Finally, the Chinese anticipate major changes in the countryside in the next decades, with an ever increasing proportion of the population moving out of agriculture into industry. Rural industry in the last five years has already shown remarkable dynamism, and this is expected to continue. However, this development requires that more advanced technologies be made available to rural industry.

These are the main objectives of current technology policy. To meet them, the Chinese have embarked on a major problem of reform of the economcy and of the R & D system. They have begun aggressively to seek foreign technology and foreign investment, and show increasing interest in participating in the international economy and scientific community. These moves mark a major change in China's S & T development strategy. To appreciate them, it is necessary to consider the legacy of China's earlier development efforts.

As China entered the post-Mao era in 1976, it had many S & T achievements behind it, particularly in the science and technology of national security. It had, largely on its own, joined the small group of nations which have acquired nuclear weapons and launched satellites. In its launch vehicle technology, China has shown an ability to apply cryogenics technology to launch vehicles, an achievement which had been made only by the US and the Europeans. Achievement was not limited to the defense area, however, as the synthesis of bovine insulin, the development of high yielding varieties of rice, and many other discoveries and inventions indicate.

China had also made considerable progress in providing for the infrastructure for R & D, and now has an extensive network of research institutes. China's R & D system of organization can be thought of as having five sectors. These include 1) the Chinese Academy of Sciences (CAS), 2) research institutes under production ministries, 3) institutions of higher education, 4) national defense and 5) research institutes under the jurisdiction of provinces or cities. Overall, these total between 9,500 and 10,000 (excluding the social sciences), with nearly 120 institutes within CAS alone, and cover most fields of science and technology.

The size, quality, and sophistication of these 9,500-10,000 institutes vary considerably. CAS institutes, for instance, range in size from those having several hundred staff, to some with more than 1,000. The Ministry of Water Resources and Electric Power, to take the case of just one production ministry, has 8 central research institutes in the electric power area (excluding water resources). These employ 5,770 persons, of whom 2,730 are engineers and technicians. One institute, the Electric Power Research Institute in Beijing, employs 880 persons. Of these, there are 477 engineers and technicians, of whom only 36 are senior engineers.[1]

The Chinese also make a distinction between "free standing" or "independent" institutes (such as those under CAS, and many of those under the production ministries), and "non-independent" institutes (such as those attached to enterprises and universities). The proportion of the latter has grown significantly since the beginning of the reform program in the early 1980s, and now account for slightly less than 50 percent of the total.

In spite of certain qualitative variation within this system of research organizations, it is nevertheless a quantitatively extensive one. Quantitative extensiveness and considerable qualitiative variation also characterizes China's pool of scientific and technical manpower. Starting from a very limited base in 1949, China's S & T manpower pool has grown to include more than 7 million S & T personnel, with close to 400,000 in R & D.[2]

Within this pool of technical manpower, there is great qualitative variation, based upon place and time of training. The Cultural Revolution unquestionably led to a qualitative decline in training, and many from the generations whose training was touched by that phenomenon are at a disadvantage today. In the period before the Cultural Revolution, however, there were elite institutions which provided sound education and training in the sciences. Those who benefitted from this training, along with Soviet trained and some western trained scientists, are serving as China's current leaders of S & T activities.

The limited access to the world scientific community during the Cultural Revolution period also affected the quality of Chinese scientific manpower. During the years of the Cultural Revolution (1966-76), there was a revolution in the technology of scientific research in the international scientific community, wrought by major advances in instrumentation. The Chinese scientific community was largely cut off from these changes, and getting access to the new technologies of research has been one of the more important benefits of the expanded student and scholar exchanges.[3]

While the achievements of Chinese policies for fostering S & T development are notable, so are the failures. Putting aside for the moment the disruptive effects on S & T of the Cultural Revolution, Chinese strategies for S & T development have for 30 years been unable to create the conditions for the effective linking of R & D and production. Thus China's often impressive laboratory achievements have often failed to lead to serial production of high quality innovative products or the employment of innovative production processes. In the words of one Chinese science policy official,

> China and Japan began their respective research on semiconductors at about the same time, and China succeeded in the experimental production of a transistorized computer before Japan did; China successfully developed a laser less than one year after the first laser came on the international scene. Today China has substantial production in its semiconductor industry, computer industry and laser industry and has tens of thousands of production workers in these industries. But we have fallen far behind international standards in these areas.[4]

Thus, since the 1950s, China's ability to achieve "intensive" economic growth through productivity gains and innovation has been disappointing. As China entered the post-Mao era, the rate of productivity growth in the industrial economy was declining, a fact which perhaps more than any others, prompted the Chinese to undertake the extensive program of economic and S & T reform we see in the 1980s.[5]

Other problems afflicted the R & D system at the beginning of the post-Mao era as well. There was often considerable duplication of research work, research productivity was low, and national direction and coordination was weak. Most importantly, The Cultural Revolution left a legacy of misassignment and mistreatment of scientists and engineers by unschooled cadres to whom they were subordinate. These various problems all point to some underlying problems with the values and organizing principles of China's economic and scientific institutions, which in turn are reflections of the evolution of the institutional order for social coordination and control in Chinese society. To fully understand the problems in the research system in the post-Mao period, therefore, we must look briefly at some of the more basic institutional problems affecting Chinese society in the aftermath of the Cultural Revolution.

Neither Market Nor Plan

As China entered the post-Mao period, it was a society of considerable institutional fragmentation. There were horizontal cleavages between central and local authorities, and vertical cleavages following the boundaries of central ministries. In addition, the work unit, or danwei, had become a significant obstacle to policy implementation. Without attempting a full explication of these phenomena, we can at least outline their origins and significance for S & T development.

Although the Chinese system of institutions shows the influence of the Soviet paradigm, it is very misleading to assume that Chinese institutions of the late 1970s were a replica of the Soviets. The hallmarks of the latter are strong centralization and "taut" planning to assure the coordination of the economy. The Chinese as early as the late 1950s were deviating from these standards. During the Great Leap Forward period of 1958-60, the Chinese began their experiments in decentralization. One enduring consequence of this first round of experimentation was to strengthen the role of local Communist Party committees, and with them, the role of local officials in personnel management.

Decentralization took on new forms in the 1960s and early 1970s with fiscal reforms which allowed local political units more powers over their finances. The right to retain depreciation funds of locally controlled enterprises in particular, gave local governments a source of revenues largely outside of central control.[6] The history of China's various experiments with decentralization has left a legacy of great institutional complexity, confusion about central and local perogatives, and at times, tension between the central government and local governments. This situation which the Chinese refer to as tiao tiao kuai kuai (literally, "branches and lumps"), greatly complicates the tasks of planning and financial management.

Effective central planning has also been frustrated over the years by the growth of influence of the large central ministries (the "branches") in China which because of their size and high degrees of vertical integration were under the best of circumstances a challenge to the coordinating intentions of central planners. Of particular importance for S & T are what had been known as the "numbered" machine building industries, all of which had much of their assets committed to national security production and research since the early 1960s.[7] These ministries were the recipients of priority allocations of money, material and manpower throughout the Maoist era.

Beginning in 1964, the Chinese began a major industrial relocation project known as the "third front construction project" (*sanxian jianshe*) which involved the movement of industry and some R & D installations from the eastern coastal provinces to interior provinces in the north and southwest largely for national security reasons.[8] This project was undertaken in secrecy, and contributed to the militarization of the industrial economy, its spatial dispersion, and its further fragmentation. Just as the "third front" program was at its height, the Cultural Revolution disrupted central planning mechanisms. This reinforced the decentralizing trends noted above, and seemingly enhanced ministerial, as opposed to, and at the expense of, central planning authority.[9]

Finally, over the years, the work unit, or *danwei*, had come to assume an enormously important role in Chinese society. The *danwei* is the basic unit of organization for the individual, providing not only employment but also housing and a variety of social services. Individuals were thus very much at the mercy of the leaders of the *danwei*, and the latter came to regard individuals working in the unit, and the products of their work, as "unit property." This attitude of "unitism" (*danwei zhuyi*) has proven to be a major problem in the post-Mao period in the face of reforms designed to achieve greater economic integration, and greater labor mobility of scientific and engineering personnel.

Thus, Chinese institutions in the beginning of the post-Mao era were somewhat different from what one might infer from a knowledge of the Soviet system. Central planning, the basis for economic coordination and control in the Soviet model, was weak. Market mechanisms, the alternative to central planning, were underdeveloped, and forms of decentralized exchange were further frustrated by the proprietary claims on resources made effectively by the *danwei*. As we shall see, this evolution of Chinese institutions had profound effects on the patterns of S & T development, and it forms a necessary background for understanding the "half-full, half-empty" record of Chinese S & T achievements. Let us consider the patterns of S & T development in greater detail by reviewing changing approaches to science policy in the post-1949 period.

Changing Approaches To Scientific And Technological Development

When the government of the People's Republic of China was founded in 1949, it very early formally expressed the view that the development of science and technology would be a central value of the new regime. That S & T has been at the center of value considerations in the development of national policy in the subsequent 30 years. Whether national policy has in fact fostered S & T is, as seen above, subject to interpretation.

This early committment to S & T can be traced back to two sources. On one hand, we can see its origins in the ideology embraced by the new regime. Marxism is of course taken by its adherents as a science, but more than that, it predicts that under conditions of socialism, modern science and technology can be expected to reach higher stages of development. These stages would not be possible under capitalism given the intense contradictions between the means of production and the relations of production which are said to exist under conditions of mature capitalism.[10]

A socialist society, therefore, was expected to be inherently more congenial to S & T development. On the other hand, pre-1949 China was not a society of advanced or mature capitalism, with highly developed means of production. It was appropriate, therefore for the new government to develop those means, and supporting S & T was one important way to do this.[11]

Quite apart from Marxist ideological influences, however, the early commitment to S & T development also had its origins in modern Chinese history. The technological and scientific superiority of the West, which made itself intensely felt beginning in the middle of the 19th century, made the issue of how to foster the development of modern S & T in China a major concern for Chinese elites, and added a nationalistic tone to the quest for modern science and technology, one still in evidence today. The old Confucian order came to be seen as a cause of Chinese scientific underdevelopment, and thus the development of science would be a force for the destruction of the backward old order. In these ways, issues of S & T development became part of the revolutionary politics and cultural ferment of early 20th century China.[12]

In addition, we see in the first three decades of the century, a series of steps taken by individuals and by the Republican government to do something about scientific underdevelopment. Chinese students began to go abroad to study science and engineering, private scientific organizations (including professional societies) were set up, and modern universities and scientific research institutes were established by the government. The latter included the Academia Sinica, which became the core of the new post-1949 Chinese Academy of Sciences (CAS). These pre-1949 developments produced the raw materials for the S & T development efforts of the new regime.

Since making that early declarative commitment to S & T, the Chinese under the People's Republic have now had 37 years of experience with S & T development. In some ways the experience seems to be representative of what other developing countries have faced. But in many important respects, the Chinese case is different. An examination of this experience suggests that there is no simple "Chinese model" for S & T development in the sense of a coherent, consistent strategy. To the extent that there is a Chinese model, it is in the way competing approaches, appearing at different times, have interacted to produce what we see in the late 1980s. It will be worth looking at these approaches in some detail.[13]

A review of China's S & T development efforts over the past 30 plus years suggests that there have been at least four different conceptions of how this development should proceed. Some clearly have been much more influential than others. I shall refer to these as: 1) the Western university model; 2) the Soviet model; 3) the mobilization model; and 4) the national security state model. The current approach of the 1980s described in more detail below, is sufficiently different from what has preceded it that it too could be called a fifth model. These will be described roughly in terms of the chronological order in which they have been important.

1. The Western University Model. In some respects this has been the least influential approach to receive formal policy consideration. It places a premium on the role of the research university as an institution for leading S & T development. The university is seen as the center of the nation's research, offering a format for combining research with education and training, and supporting the nation's modernization efforts through the supply to the labor pool of highly qualified graduates with research experience and with an appreciation for the value of research.

It was this Western University model which was advanced by China's senior scientists in the 1950s. It was what they knew best, both from their experience as students studying abroad before 1949, and in the early 1950s. It was also in many ways what best characterized the pre-1949 experience in China, in spite of the existence of the Academia Sinica. The appeals of this policy were not recognized by the political authorities in the 1950s however; it was western, it was inherently elitist in a society with strong anti-elitist values, and it was in the short run profoundly impractical about the immediate needs of the society.

The developing relationship with the Soviet Union in the early 1950s offered an alternative which it was thought at the time, would not have these problems. In addition, it was an alternative which was backed up by concrete material support in the form of Soviet assistance.

2. The Soviet Model. This became the policy of choice, and has had a profound effect on the development of Chinese science and technology. The main policy concept in terms of the organization of research, is centralization. The focus of the research system is a national academy of sciences which can command the highest quality human resources available in the society. It is the premier performer of basic research, and of much of the more sophisticated applied research as well. Industrial research, is also centralized in institutes or specialized academies serving the needs of an entire industry. In its technocratic orientation, the Soviet model was not totally inconsistent with emerging planning traditions which were developing in the pre-1949 Nationalist era.[14]

Research in universities has a decidedly secondary role in this policy. Much of the higher education sector is to be devoted to specialized training institutes, often administered by production ministries. These specialized institutions of higher education are seen as sources of the supply of manpower for the ministries and their subordinate enterprises.

This system of research, education and production is supposedly coordinated through central research planning which is to be linked to economic plans. Centralized resource mobilization and investment is to give force and effectiveness to the projects of the plan. An implicit assumption in the model is that technological innovation in production enterprises will be "supply driven" by the outputs of the R & D system and by the technologies made available to enterprises by central planners. Finally, "knowledge" in this model is regarded as being a good which has been socialized, and is freely available to those who wish it.

Early in the 1950s the Chinese decided to subscribe to the logic of the Soviet model, and set about devising institutions to emulate it. At the outset, this involved reorganizing the Academia Sinica, planning for its expansion, and introducing research planning mechanisms. At the beginning, China's supply of trained scientists and engineers was meager. Thus it was not until the second half of the decade that research in production ministries, and to some extent, in universities and other institutions of higher learning began. Often, activities in these latter sectors were spun off from the CAS, a pattern which became particularly important for the development of high-technology defense related work in the new machine building industries established to serve the military in the 1960s.

In some respects, the Soviet model served China well in the period before the Cultural Revolution. It provided mechanisms for mobilizing human and material resources and directing them to areas of high priority. As a result, Chinese science and technology expanded rapidly in the pre-Cultural Revolution years. The number of scientists and engineers expanded steadily, the number of research institutes increased, and China was able to develop a number of new fields of science and technology including those relevant to strategic weapons. Chinese scientific priorities in this initial period of Soviet influence are best reflected in the 12 year science plan first introduced in 1956. These priority areas included atomic energy, electronics, jet propulsion, automation and remote control, petroleum and scarce mineral exploration, metallurgy, fuel technology, power equipment and heavy machinery, problems relating to the harnessing of the Yangtze and Yellow rivers, chemical fertilizers and agricultural mechanization, prevention and eradication of dread diseases, and problems of basic theory in the natural sciences.[15]

There were disadvantages to the Soviet model as well. The very success of the model contained within it problems. The larger and more differentiated the system, the more it required centralized administrative attention. The relative scarcity of this administrative resource, a problem in all centrally planned economies,[16] became even more problematic as the need for it increased. This was the case in spite of the fact that the Chinese made major efforts to enhance administrative capability with the establishment of the State Science and Technology Commission (SSTC) in 1958.

The Soviet model also contains within it an imperative for bureaucratism, and organizational segmentation which works against easy, decentralized communication among different organizations. This results in a significant separation of research organizations from production organizations. When combined with the disincentives for innovation operating upon factory managers in Soviet-type systems, this segmentation produces a system in which the economic benefits from research - in the form of technological innovations contributing to productivity gains and new products - are meager relative to the investment.

Finally, the Soviet model, derived from principles of Marxism-Leninism, contains what appear to be from a comparative perspective, faulty assumptions about the nature of technology and technological innovation. Under conditions of socialism, technology is understood to be a free public good, available to all. Innovation is understood to be a "supply push" rather than a "demand pull" phenomenon. Socialist science policy, therefore, is understood to have as its main objective the production of technical knowledge through R & D in a planned fashion, and in coordination with economic plans. As noted above, the aspirations and requirements for this system to work overwhelm planners, and thus the requirements are not met, except in those high priority areas which can capture and hold planner attention. In addition, the "supply push" orientation leads to policy neglect of the demand factors - in particular, the environment for innovation (risks and incentives, prices, supplies of needed inputs and trained manpower, etc.) at the level of the production unit.[17]

Thus, the Chinese have been plagued for almost 30 years by a system in which the results of research, often not insignificant results, do not find their way into production. As early as the 1950s, Chinese political leaders began to criticize the research community for this defect. This criticism continued in varying forms and degrees of intensity for two decades until the end of the 1970s, when the leadership was finally prepared to acknowledge that the defect was in the Soviet model itself.

3. The Mobilization Model. Dissatisfaction with the Soviet model began long before the 1970s however. By the late 1950s, Mao in particular, became disturbed by some of the consequences for Chinese development of emulating the Soviet model. In the views of Mao and his supporters, the Soviet model showed some of the same signs which made the Western university model unappealing. The Soviet model was also foreign, also elitist, and as we have seen, did not contribute as fully to economic development as had been hoped. In addition, it seemed to contribute to the formation of a new technocratic power structure which was

not clearly consistent with the interests of what was still largely a rural based Communist Party. Beginning in 1958, therefore, a new approach - the mobilization model - was advanced and began to take hold. Its influence began to wane in the early 1960's, but then came back in major ways during the Cultural Revolution until it was finally rejected in the post-Mao period.

The mobilization model had a number of different elements to it. Unlike the Soviet model, it does not approach social coordination through purportedly rational planning. Instead, research and production should be approached as in a political campaign, where the object is to whip up revolutionary fervor in order to break out of established routines. By raising ideological consciousness, the requirements for social coordination will be reduced, since the properly initiated will share in a common vision to which individual efforts could be directed.

In addition, the right frame of mind and ideological orientation would provide the motivation for special efforts allowing the achievement of new breakthroughs in scientific discovery, technological innovation, and production. The mobilization model also stressed the value of self-reliance; the wisdom of the Chinese people, if properly unleashed from the psychological shackles of deference to experts, and to foreign science and technology, could become the main force for S & T development.

The mobilization model also had the objective of broadening the base of participation in scientific and technological undertakings. To overcome the drift towards elitism in the Soviet model, the mobilization model called for the involvement of workers and peasants in the tasks of research and innovation. Under the influence of the mobilization model, the history of science and technology was revised to highlight the contributions of the masses of ordinary workers and peasants to S & T development, particularly in Chinese history.

Finally, and perhaps most importantly, the mobilization model had an effect on the organization of science. Whereas the Soviet model was centralizing, and tended to concentrate R & D rsources in the hands of a central academy or central ministerial institutes, the mobilization model called for the decentralization of research, and the establishment of research and technical service facilities at provincial and local levels of government. Thus it was under the influence of the mobilization model that what some regarded as the distinctive features of the Chinese development experience - its decentralized, mass-oriented, labor absorbing, appropriate technology features - had its origins.

Assessing the consequences of the mobilization model is one of the more difficult tasks for the student of Chinese S & T development. Under its influence, there were institutional innovations which led to a useful extension of the network of S & T organizations out of the main cities. Its insistence that technical experts concern themselves with the practical problems of production, and that ordinary people not be awed by experts seems to address problems which many developing countries face. Its encouragement of innovative activities by workers and peasants would seem to be focusing attention on the problems of innovation where it belongs - in production settings, rather than the laboratory. The support for the participation of the masses in the creation of the nation's technology seem consistent with deep and worthy revolutionary political and social values.

Yet, the mobilization model has also been extremely costly to China. Its attacks on expertise were extreme and devalued the very technical knowledge which Chinese development required. Coupled with its anti-elitist themes, the attacks on expertise led to attacks on experts, and led to the tragic, gross misutilization of China's scarcest strategic resource, its technical manpower. The encouragement of worker and peasant innovation did produce innovations which served Chinese needs for a time, but they were usually crude and of no enduring value to the general upgrading of industrial technology. "Shop floor innovation" in the absence of a respect for expertise and in the context of economic institutions which did not reflect the costs and benefits of such innovations, more often than not became the "Rube Goldberg" type artifacts which were still visible in Chinese factories in the late 1970s when foreigners had the opportunities to visit. The efforts to break down the barriers between research and production by having research institutes and universities start factories, and those to overcome the barriers between organizations by coercing technical personnel into production settings, also proved counterproductive in the absence of reliable cost accounting mechanisms.

4. The National Security State Model. It can be somewhat misleading to conceive of the S & T development in the period from the middle 1950s to the middle 1960s as simply a conflict between the Soviet model and the mobilization model, although that conflict was certainly there.[18] However, there was also in this period an increasing commitment to the development of S & T

capabilities to serve national defense. In some ways, this is another manifestation of the Soviet model which also accords defense R & D a privileged position. But as the discussion below will show, there is justification for treating the national security state model as separate and distinct.

It has been know for some time that since the late 1950s, a very large share of resources, both material and especially human, were being devoted to national defense R & D. While we still do not have comprehensive figures, we are coming to see more clearly that a large share of the work of many of the institutes of the CAS (and some universities), for instance, was committed to the defense sector.[19] As research in industry expanded, with facilities and personnel often being spun out of CAS, an increasing share of this went to defense as well. This is particularly true in the machine building industry, where new ministries were created in such fields as nuclear science and technology, space and rocketry, and electronics, which were almost entirely independent of the civilian economy.

The importance of the national security state model increased in the 1960s as relations with the Soviet Union worsened, and as the conflict in Vietnam expanded. China thus came to face a largely hostile environment beyond its borders - north, south, east and west. The successful explosion of a nuclear device in 1964 was a positive feedback that important achievements were occurring in defense R & D, and that Chinese science - working largely in a self-reliant mode - could make important contributions to the national defense if it had the resources to proceed.

Faced with an increasingly hostile international environment, in the early 1960s the Chinese accelerated the program noted above to move industrial installations (and some R & D facilities) to such western provinces as Sichuan, Guizhou and Yunnan. This "third front construction" project consumed a large share of the national industrial investment for almost a decade, and thus had a major impact on Chinese industrial development at a crucial stage.[20]

As we have seen, the influence of the national security state model came at a time when many of the central coordinating and planning bodies presumed in the Soviet model, were either disbanded or greatly weakened as a result of the Cultural Revolution. The State Science and Technology Commission, for instance, was abolished in 1967. This had the effect of enhancing the discretionary powers of industrial ministries, and making them in the long run, less accountable to the center and more difficult to control. In addition, the strength of the national security state model came at a time when the military's role in Chinese politics had grown enormously as a result of the chaos of the Cultural Revolution. The combination of this political role and the allocation of so much of the nation's investment to the national defense industry, represented a significant militarization of Chinese society. This was not lost on China's key leaders, Mao and Zhou Enlai, and in the early 1970s they began what has become a long term effort to reduce the influence of the military in Chinese society.

The full significance of the influence of the national security state model on Chinese S & T development can be seen in retrospect. First, China's pattern of industrialization was sectorally and geographically determined not by the criteria of market efficiency, nor by a sense of planner efficiency, but rather by national security considerations. The control over human and material resources enjoyed by the defense sector, and its reluctance to surrender these in the new post-Mao era when national security concerns are less pressing, has been an issue in the politics of China's post-Mao "four modernizations" policy. Thus we find in the 1980s the Chinese devising new "conversion" policies for the defense sector in order to utilize the superior facilities and manpower of such industries as space, nuclear, and aeronautics to serve the civilian economy.

Finally, the relative backwardness of the civilian economy in such high technology sectors as electronics (in spite of evident possession of technological capabilities within China in these sectors, is due to the legacy of the national security state model. The organizational separation from the civilian economy, the lack of incentives to transfer technology to the latter, and the imposition of strict secrecy policies, facilitated the monopolization of these technologies by the military and prevented the diffusion of high technology to the civilian sector. In short, the problems of the separation of research from production inherent in the Soviet model were exacerbated by the adoption of the national security state model.

The achievements of the national security state model should not be overlooked however. It allowed China to develop a comprehensive nuclear industry[21] and a space industry which is now, as part of the "conversion" policies noted above, responding to the shortage of launch services internationally with offers to launch foreign satellites. In addition, China's emergence as a force in the world shipbuilding industry is again traceable back to investments made in this sector under national security state assumptions.[22] The significance of the national security state model for the future will be discussed further below.

A New Beginning

By the early 1970's, Deng Xiaoping and his supporters, backed by many western trained scientists and science policy officials were prepared to consider alternatives to the models for the development of S & T with which China had been experimenting since the 1950s. The proximate cause for this shift was the concern that Chinese science was not serving production, and that the economy was not realizing the gains in productivity that were expected of it. The openness to alternatives received powerful reinforcement following the rapprochement with the US, and the beginning of exchanges of scientists and technical delegations. Those Chinese who began travelling abroad at this time came to realize that Chinese S & T had fallen further behind world standards as a result of the disruptions of the Cultural Revolution than they had realized. These impressions were reported back to China's highest leaders, who also were meeting frequently with Western scientists who were visiting China.

Until Mao's death in 1976, and the subsequent purging of his radical followers, not much happened. However, as the post-Mao leadership finally began to coalesce by the end of the decade, there was an explosion of interest in Western science and technology, and China initiated a new foreign policy - the "open door" - to try to capture this foreign S & T for China's benefit. But along with the open door came a serious, and systematically pursued interest in how science and technology was organized and implemented in the West. The superiority of Western science, and especially technology, in comparison with Soviet S & T was obvious in ways that were not the case in the 1950s.

As the Chinese studied the Western experience, they began to see a system for the production and utilization of knowledge that was different than what they had formerly believed about the Western university model, and could see that at least some of this system had applicability to China. In China's "new" perceptions of the Western system, the research university does play an important role, as in the earlier Western university model. However, it is a role played in a context which is also defined by widespread industrial R & D, and by government policies using, in combination, both direct (eg. research grants) and indirect (eg. tax policy) means to support research and innovation, which were far more sophisticated than anything the Chinese had ever tried. They also came to realize that in the Western approach to technological innovation, factors affecting the demand for new technology are as important, if not more important that those effecting supply, which are given prominence in the Soviet model.

Thus, under the combined influences of dissatisfaction with their own past experiences with the shifting, interactive consequences of the Soviet, mobilizational, and national security state models, and impressed by what the West had achieved following very different approaches, the Chinese set about in the early 1980s to reform their economic and R & D systems in unprecedented ways. It would be tempting to refer to this reform effort as the introduction of the "revised Western model." However, the Chinese realize that there clearly are limits to the wholesale emulation of the West, given the underlying socialist principles of the economy.

The main features of the current reform program, reflecting this Western model are as follows:

1. The relationship between central decision making and the individual research center has changed. The Chinese want to achieve both more effective central decision making, and more initative from the grass roots. A major change to strengthen central coordination has been the establishment in 1982 of the supra-ministerial Science and Technology Leading Group. The latter is directly subordinate to the State Council and its nominal head is Premier Zhao Ziyang. At the same time, individual research units are being given more autonomy with reference to funding, personnel and the choice of research topics. Thus, there is a simultaneous centralization and decentralization going on.

2. Issues of centralization and decentralization are also evident in the organization of R & D. The role of the CAS is being given intense scrutiny. There is a good possibility that it will undergo major changes in the next five years, perhaps by divesting itself of some of its nearly 120 institutes, and having others converted into something which would be more like the national laboratories of the US or Western Europe. Similarly, there is reorganization in the area of industrial research. The idea of centralized ministerial institutes is giving way to a more differentiated system of facilities. Some of the former ministerial institutes are being converted into contract research centers, while others are being absorbed

into industrial corporations as in house corporate research laboratories (a reflection of the growth of "non-independent" research centers noted above).

3. Major changes are being made in the funding for research. A significant break from the Soviet approach of block funding to institutes through an annual budget is being attempted. The influence of the West is directly evident in the initiation, in early 1986, of a Chinese National Science Foundation for the support of basic research, and the introduction of competitive bidding and contract research in the area of applied research. The underlying objective in these changes is to introduce an element of competition to the R & D system, and to make it more cost effective and accountable. The Chinese have also established their first venture capital company in an effort to experiment with new forms of financing high technology industry.[23]

4. A fourth area of reform is the allocation of technical manpower. The Chinese practice in the past was to allocate personnel through administrative assignment, in keeping with an underlying principle that socially optimal allocations of goods could be realized through rational planning. This assumption has increasingly been questioned in the face of much evidence that manpower is being terribly misused. In addition, the exposure to Western practice has indicated that the kind of dynamic technical change they desire in their economy is not possible without opportunities for technical people, the carriers of technology, to move about and change jobs. The Chinese are therefore experimenting with a variety of labor market-type mechanisms which they hope will add flexibility to the manpower allocation process. Reform of the personnel system is among the most politically sensitive goals and change is likely to come slowly.

5. The current reform thinking assigns to universities a very different role in the research system than one finds in the Soviet model. Following Western practice, especially selected universities with superior faculties and traditions of quality, are now seen as important centers of research. Efforts have been made to upgrade the facilities (in part, through assistance from the World Bank) for university research and to improve the climate for it. An additional manifestation of this change is the introduction of advanced degree programs for the awarding of masters and doctoral degrees at universities (and some research institutes) where the faculties have been certified as qualified to supervise graduate programs. Contrary to Western practice, however,

there is intense pressure on the universities to orient their work towards applications, and some Chinese and foreign observers believe that this policy will thwart the development of a productive university research tradition.

6. A much touted area of reform is the introduction of what the Chinese call "technology markets." These take various forms, including the convening of special fairs for the exchange of technology. The opening of technology markets seemingly has been an important stimulus to internal technology transfer. According to one report, the value of domestic technology transfers in 1985 was 2.3 billion *yuan* (US $ 720 million), a 300 percent increase over 1984.[24]

The real significance of the technology markets, however, is symbolic. They mark a change in attitude and philosophy about technology, and represent a break with the assumption of the Soviet model that technology is a free public good. Instead, it is being regarded as a commodity which can be bought and sold. In addition, the Chinese have come to realize that there are issues of property rights attending to technology, and have introduced a patent system to deal with these.

7. A final component of the reform program is the attempt to reorient the work of the defense establishment to serve civilian industry. Again we see a shift from a Soviet style military industrial complex with strict separations of the military from the civilian, to more of a Western style, in which defense plants have both military and civilian customers. With this reform, the Chinese hope to capture the resources in the defense sector for socially more productive purposes in ways not found under the assumptions of the national security state model. It is also true, however, that another motive for the involvement of the defense establishment in a civilian economy which is becoming increasingly market driven (or "marketized") is to make defense R & D and production more efficient by subjecting it to the forces of competition.

As noted above, the national security state model, while terribly wasteful, nevertheless permitted the Chinese to mobilize the resources needed to launch a series of strategic science-based industries in such fields as space, nuclear, computers and electronics, and materials. Ironically, however, these industries have not been able (with a few exceptions) to achieve world standards of research and production. However, they do represent a comprehensive S & T infrastructure which, under the right conditions, could be able to move rapidly into world markets with advanced products of quality. These conditions include the availability of proper incentives and capable management, and access to advanced foreign technology.

The latter has become a focus of Chinese interest in recent years in many industries, including the defense production ministries. The latter have undergone considerable reorganization, along with the name changes noted above, and have all established their own foreign trade companies to both market their products internationally, and to acquire technology. Whether these industries, with their more advanced technological infrastructure, can effectively assimilate foreign technology remains to be seen. The case of the shipbuilding industry, which was one of the first to undergo reorganization and to alter its mission, does seem instructive. Its growing importance as a source of international supply is attributable not only to its cost advantages, but also to its ability to incorporate advanced foreign technology into an already well established industry, with strong research and design capabilities.[25]

China's current reform program is appropriate in that it offers China an opportunity to realize more fully the practical benefits from its investments in S & T development which have eluded it in the past. The Chinese see important links between the reforms in S & T and the more general "marketizing" reform program for the economy as a whole. Both programs are in the abstract, well conceived, and are intended to alleviate the severe problems of the economic and S & T institutions noted above.

There are a number of issues concerning the future prospects for reform, a full discussion of which is beyond the scope of this paper. The reform programs are complex, have encountered resistance already from some Party cadres, managers and scientists who prefer the comforts of a more subsidized existence, and are likely to encounter more. They have also produced unexpected consequences, such as the undervaluing of basic research in the rush to promote contract research, and are likely to produce more in the future. Effective political leadership will be required to stay the course of the current reform program, as will patience. China's leaders seem to realize that the full working out of the reforms will take at least five years.

The Experience With Technology Transfer

In addition to the efforts to develop its own indigenous technical capabilities described above, the

Chinese search for a strategy for scientific and technological development has also included the transfer of foreign technology to China. From the early 1950s to the late 1970s (excluding the surge in imports which began in 1978) the Chinese expended approximately $12.9 billion on technology. This was a relatively small amount in relation to the size of the economy, but a very high percentage (93) of this expenditure during this period went for whole plants and equipment.[26]

Chinese interest in technology transfer has varied over time, however, and Chinese attitudes towards foreign technology must be seen against the background of Chinese modern history. Since the late 19th century, the Chinese have realized the importance of modern technology from the west, but have been unable to reach a culturally congenial relationship to it. As characterized by one foreign trade official who was trying to dispel these attitudes, Chinese thinking often goes as follows:

> "After the Opium War, imperialist powers carved up China. Our political and economic lifelines were controlled by foreigners, our markets were flooded with foreign goods, and our national industry was severely devastated. Such a period of national humiliation is still fresh in our memory. Therefore people always associate imports with the protection of our national industry, and tend to think that the less imported the better."[27]

Not surprisingly, therefore, the Chinese have been concerned that importing of foreign technology not lead to dependency, an over-reliance on foreign material culture, and the betrayal of Chinese culture. The salience of this historic concern was reinforced at the end of the 1950s, when after allowing themselves to become quite dependent on Soviet technology during the 1950s, the Sino-Soviet relationship soured, and Soviet technical assistance was withdrawn.

China's experience with importing technology since 1949 can be divided into four stages. The first was from 1950-1960, when China imported technology from the Soviet Union and Eastern Europe in support of 156 major industrial projects focusing upon such basic industries as metallurgy, machine building, trucks, coal mining, electric power and petroleum. Some 400 items of technology were introduced, with an approximate value of $2.66 billion. These transfers were indispensable for the timely establishment of new industries, and contributed to the rapid economic growth experienced at the time.[28]

The second period was from the time of the withdrawal of Soviet assistance in 1960, to the outbreak of the Cultural Revolution in 1966. During this period, the Chinese began relying more on Japan and Western Europe for technology. Some 84 major contracts, worth $280 million were signed in this period. Industries targeted were metallurgy, chemicals and chemical fibers, and synthetic textiles.

The third period extends from the early 1970s to 1978. In the years from 1973 to 1978, China signed some 300 contracts for foreign technology worth $9.9 billion. The emphasis in this period was on complete plants in such industries as steel, petro-chemicals and chemical fibers. Many of the contracts from this period were concluded in great haste in late 1978 and subsequently were cancelled or postponed.[29]

It is interesting to note that following China's technology transfer experience with the Soviet Union, which was "intimate" in that it involved a whole range of transfer experiences (including the importation of whole plants, the supply of Soviet blueprints to the Chinese, the presence of Soviet technical advisors in China, and the training of Chinese in the Soviet Union), China's subsequent approaches to technology acquisition in the 1960's and 1970's were more "arms length," focusing on the importation of complete plants and sets of equipment without due attention given to the "software," training, and advisory services which often contribute to successful assimilation.

The fourth period is from 1979 to the present. A number of changes in China's approach to technology transfer have been made during this period. By 1979, the Chinese came to the conclusion that the arms length mode of transfer focusing on complete plants was both too costly and did not yield the know-how they expected. Since then, Chinese policy has discouraged the acquisition of complete plants and equipment, and has stressed the acquistion of know how; "acquiring the hen, and not just the egg," as the Chinese put it. Thus, modes of technology transfer which offer more intimate interactions with foreign technical personnel have come to be preferred. A wide variety of instruments of transfer, including licensing, joint ventures, cooperative ventures, wholly foreign owned ventures, compensation trade, and the use of consultants and the procurement of technical services are being used. Much emphasis is being placed on foreign provision of training in Sino-foreign technology transfer contract negotiation. As a result of this change, as shown below, a much greater proportion of the technology imported since the end of the 1970s has been "unembodied" technology, or pure know-how.[30]

In addition, China has been spending a greater percentage of its resources on importing technology than it did in the past. In the Sixth Five Year Plan period, for instance, $9.7 billion, or 15 percent of the investment funds provided for in the plan went for foreign technology.[31] Two other changes are notable in this fourth period of importing technology. Whereas in the past, the emphasis had been on technologies supporting the establishment of new enterprises, since the early 1980s, the emphasis has been on technologies to be used in upgrading or renovating existing enterprises. Finally, there has been a change in the locus of decision making on technology transfer. As part of the reformist decentralizations, it is no longer the central ministries and a single foreign trade corporation which are the principal decision makers. Instead, many other players have become active, including enterprises, local governments and a myriad of new trading corporations.

Thus in this current fourth phase of the PRC's technology transfer experience, the absolute amount of technology imported has increased notably, with a total value in the 1979-1985 period of some $9.7 billion. The Chinese also continue to rely primarily on the OECD countries for technology, and have increasingly emphasized the importation of know-how rather than technology embodied in plants and machinery.

Enduring Tensions In China's Development Experience

As Chinese programs to develop indigenous science and technology capabilities and to import technology evolve in the future, they will be influenced by two enduring issues which have characterized China's S & T development in the past.

1. The "Zhong/Wai" Tension

There is in modern Chinese history, as we have seen, an underlying tension between that which is foreign (*wai*) and that which is Chinese (*zhong*). This tension has manifested itself in the difficulties China has had in reaching an accomodation with Western material and intellectual culture, and thus pertains directly to S & T development as well.

The *zhong/wai* tension clearly has a policy

manifestation in such policy areas as self-reliance, efforts at import substitution, or more generally, the current policy of the "open door." But its significance is also cultural and psychological. At this level, one sees manifestations of the tension in what is at certain times uncritical fawning over Western science and technology, followed by periods of counter-reaction in which interest in or admiration of Western ways is seen as somehow "un-Chinese," or a betrayal of the country (the nationalist critique) or a betrayal of the interests of the broad masses of workers and peasants (the Marxist-Leninist critique).

The cultural and psychological dimensions of the zhong/wai tension also contribute to an underlying "technological nationalism" in Chinese development experience. Technological nationalism leads to the belief that China can accomplish anything that can be accomplished by Western science and technology if she puts her mind to it. Chinese achievements in space technology, nuclear weapons and the synthesis of insulin all have been strongly influenced by this technological nationalism.[32] In its more positive manifestations, it represents the kind of healthy self-reliance and self-confidence which other developing countries could use. In its less positive manifestations, it leads China to actions which are neither economically nor technologically sound.

The zhong/wai tension is also evident in China's approaches to the acquisition of foreign technology. In spite of the many benefits to China from transfers of technology from the Soviet Union, for instance, when Soviet assistance was terminated at the end of the 1950's, the Chinese came to feel that in many ways Soviet assistance was a mistake. It had produced an unwanted dependence on foreign technology, and distorted China's development.

In spite of the new orientation towards foreign technology we see in the 1980s, and the fact that Chinese conditions today are very different from what they were in the early 1960s, signs of the zhong/wai tension continue. These range from concerns about the corrupting influences of the West, to the feeling that China runs the risk of becoming technologically dependent on the West by sacrificing its own R & D and "infant industries" in the quest for western technology. Careful observers of China's decision making regarding Western technology during the past few years see the forces of technological nationalism at work as Chinese domestic suppliers of technology struggle with user industries (who often prefer foreign technology) over the question of just how "open" China's "open door" should be.

2. Economic vs. Technocratic Orientations

A second underlying them in China's S & T development is the tension between economic criteria and technocratic criteria in the making of decisions concerning S & T development. The former pertain to the application of S & T to economic development. They focus on the contributions of S & T to economic growth, seeing S & T basically as economic factors. The relevant economic theory being applied may vary; it could be some version of market efficiency, as has been the case in the 1980s, or it might be an economic approach derived from a planning procedure, as was the case in the 1950s.

In the technocratic approach, S & T development is seen more as a good in and of itself, and also as a good which contributes to control and power. The control may be more effective control over nature, or perhaps control over other organizations or countries (in the military sphere) made possible by the acquisition of a technology. The technocratic approach to decision criteria, in short, is driven by different expectations than the economic approach.

One would therefore expect to find in Chinese development experience that these different criteria have had behavioral consequences, with different actors approaching S & T decision making with different values. Thus not all bureaucrats or economic planners will necessarily embrace the same criterion. The more oriented one is to the achievement of economic results, the more partiality for the economic criterion. On the other hand, other planners and bureaucrats clearly have stressed the technocratic value of control. In this they find ready allies with some scientists, who are more concerned with S & T development for its own sake.

This latter alliance emerged in the late 1950s, and characterized the Chinese S & T system in the early 1960s before the onslaught of the Cultural Revolution.[33] In the attacks on S & T during the Cultural Revolution, they were indeed directed at both parties in this alliance, the scientists and their technocratic patrons in the R & D bureaucracies.

In the post-Mao era, especially since the early 1980s, Chinese policy-making for science and technology has become much more pluralistic and open to policy analysis than it was in the past. Scientists have opportunities to express their interests through the Chinese Association for Science and Technology (CAST), the peak organization for the nation's professional societies, through a variety of government advisory bodies, and through an Academy of Sciences which enjoys far more autonomy from political controls than at any time in its history. The SSTC has become an important advocate for the interests of the scientific community as well as a source of ideas for science policy. The SSTC is also working more closely with the Planning and Economic Commissions to effect better coordination between R & D and the economy at the macro level. Finally, the Science and Technology Leading Group, at the State Council level, attempts to integrate the diverse values of scientists, economists, planners and the military into coherent national policy.

Nevertheless, the tension between the economic and the technocratic orientations is still in evidence.[34] Those science bureaucrats who have been most influential in devising and promoting the reform program are clearly much more influenced by economic criteria. Reforms in S & T, especially reforms in the funding of R & D, are designed to make the latter much more cost efficient, even at the expense of long-term research protocols which, if evaluated on non-economic grounds, might hold up quite well. Not surprisingly, therefore, scientists are not uniformly behind the reforms, even though in many other ways, they are intended to improve their lot as an occupational group.

On the other hand, the influence of the technocrats is also in evidence. The many reports of Chinese interest in advanced technology for its own sake, all reflect the technocratic orientation. The prominence and high priority given the space and nuclear industries suggests that there are some who continue to see the preferred path to China's technological development in the "macro-management" of big national projects, through top-down "command" decisions on technological choice, in contrast to the more market-oriented approach of those embracing the economic criteria.[35] What is new is that the changes in policy making noted above, have provided a more institutionalized arena for managing the tension.

These two tensions - the zhong/wai and the economic/ technocratic - are likely to continue to affect Chinese S & T development for the remainder of the 20th century. Although there is no necessary reason that these tensions will become destablizing (all countries have some version of them), how they are managed will determine the politics of science in China. Since the politics of science has had such a profound influence on S & T development in the past, it is likely that the direction of politics in the future will be the single most important factor in whether or not

China is able to achieve the goals, noted at the outset, it has set for itself.

Conclusion

As we have tried to show in the above, there is no single "Chinese model" for the development of S & T. To an extent that seems to be unique among developing countries, China has changed approaches. It has experimented with an approach which served a useful purpose in rapidly facilitating the establishment of new sciences and technologies, in support of new industries. It has experimented with one which offered opportunities for mass participation in innovative activities and which helped redistribute S & T resources away from the more advanced part of the society to the less advanced. It has experienced what is involved in implementing policies which drive S & T development through a concern for national security.

China's experimentation has been costly. It has wasted resources and has been disruptive of a process which inherently requires a degree of policy stability over the long term. Yet, the experimentation has also had value. The pursuit of S & T development following different models has helped China achieve a number of its modernization goals.

While China's experimentation with different strategies for S & T development has been varied, it is useful to conclude with a reflection on the wider comparative context. Because of the socialist nature of the economy and the central concern for S & T development articulated by the new regime as early as 1949, scientific and technological development since the establishment of the People's Republic has been a major governmental *policy* interest, and has elicited a "top-down" programatic response from the state. This has been true for all the models followed.

Thus, China belongs in that category of developing countries which have approached technological development through conscious "science" or "research" policies. This is in contrast to those countries (such as the Asian NIC's) whose policy interventions in support of technological development have focused more on the macroeconomic climate for acquiring technology from abroad, stimulating exports, and particularly in the case of Japan, choosing strategic technologies for industries targeted for development.

The Chinese in the post-Mao period have studied the

experiences of the countries in this latter category, and clearly are attempting to emulate some of their practices, as the discussion of technology transfer above suggests. China is doing this, however, with established industries and research institutions which were founded upon very different assumptions. As we face a future in which China seeks to capitalize on foreign technology and realize the benefits of competition in the international economy in emulation of the NIC's, the role of these established industrial and S & T structures becomes an intriguing question. Will they be a springboard for rapid S & T advance, or will they be conservative forces anchoring China in a tradition of disappointment?

NOTES

1. Wang Huijong, "Some Aspects of the Science and Technology System in Modern China," Paper presented at the annual meeting of the American Association for the Advancement of Science, Philadelphia, May, 1986.

The figures provided by Wang illustrate two problematic features of employment in Chinese research and development. First, the number of non-technical personnel is often a very large percentage of the total employed by an institute. Second, the numbers considered to be "senior" scientists or engineers are a small percentage of the total, a fact which is best explained by the impact of the Cultural Revoluation on science and education.

2. Leo A. Orleans, The Training and Utilization of Scientific and Engineering Manpower in the People's Republic of China, U.S. House of Representatives, Committee on Science and Technology, October 1983; and "Graduates of Chinese Universities: Adjusting the Total," The China Quarterly (forthcoming).

3. This issue is explained in Richard P. Suttmeier, "Academic Exchange: Values and Expectations in Science and Engineering," Paper presented to the conference on Sino-American Educational and Cultural Exchange, the East-West Center, Honolulu, Hawaii, February 18-22, 1985.

4. Luo Wei, "The Position and Role of The Academy of Sciences in the Chinese Research System," Ziran Bianzhengfa Tongxun (Journal of the Dialectics of Nature) 3, No. 3 (June 10, 1981). In Joint Publications Research Service 81620, p. 90.

5. See Elizabeth J. Perry and Christine Wong, eds.,

The Political Economy of Reform in China, Council on East Asian Studies, Harvard University, 1985, "Introduction" and Gene Tidrick, Productivity Growth and Technological Change in Chinese Industry, World Bank Staff Working Papers, Number 761, Washington, DC, 1986.

6. Barry Naughton, "The Decline of Central Control Over Investment in Post-Mao China," in M. D. Lampton, ed. Policy Implementation in Post-Mao China, University of California Press, forthcoming.

7. These ministries, which had been known as the 1st, 2nd, 3rd..etc Ministries of Machine Building" were renamed the Ministries of Electronics, of Ordnance, of Aeronautics, of the Space Industry, of the Nuclear Industry, in the early 1980s, and given civilian as well as military missions. The former 6th Ministry of Machine Building has been renamed the China State Shipbuilding Corporation.

8. I am indebted to Nobuo Maruyama for calling my attention to the importance of the "third front" program.

9. There is still a great deal which is not known about the "third front." A useful recent discussion is found in Nina Phyllis Halpern, Economic Specialists and the Making of Chinese Economic Policy, 1955-1983, Ph.D. dissertation, University of Michigan, 1985. Halpern's analysis stresses the reinforcement of decentralizing tendencies and the weakening of central planning, without addressing the issue of enhanced ministerial power.

10. For example, Liu Chun and Ming Tinghua, "Science and Technology Are a Magic Weapon for Building Socialism," Renmin Ribao, July 16, 1982.

11. The ideological foundation for China's socialist development strategy is complex and beyond the scope of this chapter. When concrete problems of development policy faced the Communist leadership later in the 1950s, ideology proved to be an uncertain guide to action, and major differences in interpreting ideology appeared. These led to major divisions within the leadership between those who followed Mao Zedong's interpretations and those who did not.

12. Danny Wynn Ye Kwok, Scientism in Chinese Thought, 1900-1950 (New Haven: Yale, 1965).

13. The term "model" as used here includes the idea that there are certain organizing principles, values and operating codes for the scientific enterprise which state or assume objectives and the means to achieve them. My meaning is close to the idea of different conceptions of how science and technology should be institutionalized.

14. William C. Kirby, "Technocratic Organization and Technological Development in China, the Nationalist

Experience and Legacy, 1928-1953," paper presented at the Conference on China's New Technological Revolution, Center for East Asian Research, Harvard University, May 1986.

15. For a discussion of the 12 year plan see Richard P. Suttmeier, Research and Revolution (Lexington, Mass.: Lexington Books, 1974), pp. 58-61.

16. This point is made nicely in R. Amann and J. M. Cooper, Industrial Innovation in the Soviet Union (New Haven: Yale, 1982), Chapter 1.

17. Kazimierz Poznanski, The Environment for Technological Change in Centrally Planned Economies, World Bank Staff Working Papers, Number 718, Washington, DC, 1985 provides a useful synthesis of what is known about the defects of Soviet style systems for promoting research.

18. Suttmeier, op. cit.

19. As much as 80 percent of the work in some institutues, such as the Institute of Metallurgy, was defense-related in the pre-Cultural Revolution period.

20. The Chinese now talk about as many as 2000 factories and research institutes which were part of the "third front" policy. Beijing Review, December 12, 1985. It has been estimated that as much as 50 percent of industrial investment for almost a decade went to "third front" construction.

21. Richard P. Suttmeier, "China's Nuclear Power Option," in U.S. Congress, Joint Economic Committee, China's Economy Looks Toward The Year 2000, Washington, DC, 1986, pp. 87-103.

22. Li Rongxia, "China's Shipbuilding: Good Days Ahead," Beijing Review, June 23, 1986, pp. 17-23.

23. See for instance, China Daily, June 16, 1986, p. 4.

24. Chen Shenyi, "The Domestic Transfer Mechanisms in China." Paper presented at the annual meeting of the American Association for the Advancement of Science, Philadelphia, May, 1986.

25. Li Rongxia, "China's Shipbuilding..", op. cit.

26. Huang Fangyi, "Analysis and Suggestions of China's Introduction of Foreign Technology and External Trade," Asian Survey, forthcoming.

27. Wei Yuming on "Open Door Policy, Trade with Japan," Xinhua, October 25, 1985 translated in Foreign Broadcast Information Service (FBIS), October 29, 1985, a4.

28. Ibid. Also Robert F. Dernberger, "Economic Development and Modernization in Contemporary China: The Attempt to Limit Dependence on the Transfer of Modern Industrial Technology From Abroad," in Frederic Fleron, ed., Technology and Communist Culture, The Socio-Cultural Impact

of Technology under Socialism (New York: Praeger, 1977).

29. Huang Fangyi, "Analysis and Suggestions," op. cit.

30. Ibid.

31. Ibid.

32. A recent manifestation of this nationalistic pride in indigenous achievements is a spate of articles in celebration of Deng Jiaxian, said to be the father of Chinese nuclear weapons. While these articles can also be seen as a defense of the contributions of China's often maligned scientists in the face of lingering anti-intellectualism, the theme of technological nationalism is also much in evidence. See, for instance, Gu Mainan, "Deng Jiaxian, A Man of Great Merit in Developing Two Bombs," Liaowang 25, June 23, 1986 translated in Foreign Broadcast Information Service, Daily Report: China, July 14, 1986, K13 ff.

33. Suttmeier, Research and Revolution, op. cit.

34. One of the more dramatic indications of this tension is the reported difference between two of China's current vice-premiers, both of whom are likely candidates for the premiership, over the future of the Chinese civilian nuclear program. Li Peng, representing technocratic interests, is reported to be pushing for the implementation of the current program, while Tian Jiyun, representing the economic orientation, is said to be cool to the nuclear program in light of cost and environmental considerations. These differences are also evident in the policy debate surrounding the mammoth Three Gorges dam project on the Yangtze River. South China Morning Post, July 14 and 15, 1986 translated in Foreign Broadcast Information Service, Daily Report: China, July 16 and 17, 1986.

35. Denis Fred Simon, "The Evolving Role of Reform in China's Science and Technology System: A Critical Assessment," paper delivered at the 15th Sino-American Conference on Mainland China, Taipei, June 8-14, 1986. See also Richard P. Suttmeier, "Moon-Ghetto Problems in China's Alternative Scientific Futures," in Norton Ginsburg and Bernard A. Lalor, eds., China: The 80s Era (Boulder: Westview, 1984) pp. 303-325.

8

India: Success and Failure

Ward Morehouse and Brijen Gupta

Since 1929, the year when Jawaharlal Nehru presided at the annual convention of the Indian National Congress, India's body politic has been dominated by the political and economic ideas advanced by Mahatma Gandhi and Jawaharlal Nehru. An uneasy coalition has existed between the political systems advocated by these two makers of modern India. The development of Indian capacity in science and technology especially during the four decades since independence (1947) has had two basic objectives that derive directly from the basic thinking of Gandhi and Nehru. The first has been to meet the challenge of the rising expectation of the Indian people in their most fundamental material and social needs such as food, shelter, health, learning, and work. The other has been to eliminate -- or at least, diminish -- the dependent ("colonial" or "satellite") industrial and technological relationships with the advanced countries in the North and thus to assert greater economic and political autonomy in the international system. For much of these four decades, this firm aim was to be accomplished, at least as a matter of rhetorical disposition, by creating a socialist pattern of society, infused with Gandhian ideals.

Much to its sorrow, India has discovered that science and technology appropriate for the first end are not the appropriate science and technology to achieve the second objective. Consequently the policies that have shaped capacity-building in Indian science and technology have swung like a pendulum and have not yet found their moorings. It can be argued, as we do in this chapter, that India has had much success in creating a "modern" sector of its economy and an infrastructure in advanced technology that enhances national political power. But it also can be argued that India has failed to any significant degree in creating a technological base appropriate to its size and factor endowments and that it has also failed in meeting the most vital needs of a vast majority of its population.

Whatever may be its failures, India has, among Third World countries in the post-World War II era, succeeded to an unusual degree in institutionalizing the capabilities in modern science and technology that it set out to develop with the advent of independence. Yet the most critical question remains: Is this enough to enable India to "leapfrog" into the twenty-first century, as Rajiv Gahdhi, the current Prime Minister and the grandson of the architect of this strategy, Jawaharlal Nehru, insists is the critical national task for the balance of this century?

Indigenous Origins Of Modern Science and Technology

India did not, of course, start from scratch in building its scientific and technological capacity when it became independent in 1947. Understanding of what came before, however, is clouded by myth and conflict.

We do not wish to minimize the lively debate that has existed in recent decades between and among the historians of science on the contributions of Indian (and other Asian) science to modern (basically European) science, and the condition and course of Indian (and other Asian) science since the sixteenth century.[1] The debate centers on whether Indian (and other Asian) science and technology were hindered or promoted by the Western political and intellectual domination of India (and other parts of the world)? It can be firmly stated that between 1850 and 1950, a re-oriented, European-value dominated, science and technology slowly emerged in India (and other parts of Asia). George Basalla calls this the period of colonial science and technology, a period which saw the gradual emergence of scientists, technologists, and technicians either formally trained by Europeans, or informally schooled by works of Europeans, and the establishment of institutions and laboratores with instruments and supplies produced and distributed by European manufacturers.[2] Furthermore, during this period of colonial science, Indian scientists generally investigated problems that were delineated by European scientists.

By 1947, India had a handful of distinguished world-class scholars, including 1930 Nobel Laureate physicist C. V. Raman, and a modest scientific community. In addition to colleges and universities providing education in liberal arts, there existed in India a number of organizations dedicated to the promotion of science and technology such as the Indian Association for the Cultivation of Science (founded 1876), the Institution of Engineers (founded 1920), the National Academy of Sciences (founded 1931), the National Institutes of Sciences (founded 1935), and several disciplinary professional organizations in mathematics, physics, chemistry, agriculture, medicine, zoology, and the like. Fundamental and applied research of international distinction was being carried out at such institutions as the Forest Research Institute, the Indian Agricultural Research Institute, and the Indian Institute of Science.

Yet the size of this scientific community, compared to a large country like India (then soon to be divided into India and Pakistan), was quite small. Then one compares the progress of science and technology in India during the first half of the twentieth century with similar progress in Australia and New Zealand, the achievements in India appear deeply disappointing. Even more disturbing were the facts that polytechnics dedicated to the training of paraprofessionals and technicans never really got off the ground before 1947. True, almost every province of the country had an engineering college: but the total number of students in these colleges in 1946 was less than 7,000. Scientific modernity had yet another enemy. A section of Indian nationalists made unfortunate and obscurantist appeals to India's own premodern but venerable tradition in science and technology. Such compensatory pride in an irrelevant scientific past was hardly conducive to the growth of modern science and technology.

Prior to 1947, India's principal experience with technology lay in railroads and irrigation and to a lesser extent in the building of textile and sugar industries. Even that experience, barring a few exceptions, was in the management of European technology, not with the development of autonomous capabilities. Yet thanks to Indian nationalism, to World War II, and to a continuous tradition of educational reform since 1900, India in 1947 was mentally poised for a rapid take off in science and technology though it remained to be determined whether this science and technology was to serve primarily the Gandhian end of lifting India's rural masses to a decent standard of living or to the Nehruvian preference for creating in India a world-class science and technology order sufficiently competitive with at least most east and southern European nations.

Government And Science: Instrumentality And Independence

To grasp the essentially instrumental character of the government-science relationship in India since independence, it is necessary to examine the perceptions of science and science-based technology as forces in the contemporary world held by India's political leaders. Two leaders are most important in this regard: Gandhi and Nehru. Both opposed British rule because it was foreign. Both realized that unless Indians could be roused to life and made conscious of belonging to one large society no movement against the British would succeed. Therefore they both opposed the old fragmentations of society -- castes and communal religious groups -- and mounted special attacks upon the political economy that the British had created in India. But here the basic similarity ended.

With uncompromising force, Gandhi opposed what he called "mimic anglicism", that is western clothes, western bourgeois life, western egalitarianism, and western desire to get wealthy and to improve one's standard of living indefinitely. And so he preached that the ultimate aim of Indian nationalism was *Ramrajya*, the revival of the ancient self-contained village community, where people would remain poor, but not in poverty, because they would not aspire to consuming more than they could produce in the village or obtain in barter from the next village. Said Gandhi:

> I would say that if the village perishes, India will perish too. India will no more be India. Her own mission in the world would get lost... We have to concentrate on the village being self-contained, manufacturing *only for use*. Provided this character of the village industry is maintained, there would be no objection to villagers using even the modern machines and tools *that they can make and can afford to use*. Only those should not be used as the means of exploitation of others.[3] (Original italics.)

These basic ideas were clarified and given concrete shapes in the writings of J. C. Kumarappa (1948) and Bhartan Kumarappa (1965).[4] They provided the foundation for what is called the appropriate technology movement in India which emphasizes labor-intensive, employment-oriented, and resource-protective technologies consistent with Indian factor endowments. With Gandhi's death in 1948 and the pre-eminence of Nehru at that time, these Gandhian ideas were removed from the arena of political debate.

Nehru regarded science as an important factor for change in social attitudes, values, and outlook through the widespread dissemination and inculcation of what he called the "scientific temper." He was, however, mainly concerned with the application of modern scientific knowledge through technology as a critical means of achieving rapid material progress in a poor society and as a source of national power for a newly independent country. Already in an angry letter to Mahatma Gandhi in 1945 he had firmly stated that in order to protect India from foreign aggression and to achieve economic independence, India had to be made "a technically advanced country."[5] He felt that "a heavy engineering and machine making industry, scientific research institutes, and electric power" were the three fundamental requirements of India's independence.[6] Right from 1929, when in his presidential address to the Indian National Congress he had declared complete independence of India from Britain without any Dominion Status, Nehru, perhaps under the influence of Lenin, had set his mind in favor of modern science and contemporary technology. "It is science alone that can solve the problem of hunger and poverty, of insanitation and illiteracy, of superstition and of deadening custom and tradition, of vast resources running to waste, of a rich country inhabited by starving people," Nehru once observed. "Who indeed can affort to ignore science today?...The future belongs to science and to those who make friends with science."[7]

With regard to technology, Nehru asserted that "it is technology which has made other countries wealthy and prosperous, and it is only through the growth of technology that we shall become a wealthy and prosperous nation."[8]

Whether he was aware of the complexities of the process of linking scientific research through technology to the productive sectors of society is less clear. Nehru's role in developing India's scientific capabilities still cannot be assessed definitively, but what evidence we have suggests that he assumed cultivation of India's scientific capabilities as a sufficient public policy goal in and of itself. He believed that once these capabilities were sufficiently strong, the critical linking process through science-based technology would occur on a broad scale. This would bring about, he also believed, the transformation of Indian society along economically and socially humane lines and the acquisition of the economic and political power essential to a modern nation state.[9]

Nehru's veiws about the role of modern science and

science-based technology in India's development as an independent nation are important because of the central place which he occupied in the first decade and a half of Indian independence. His ideas and views tended to dominate public policy for science and technology in India in the 1950s and 1960s. These policies were expressed in a variety of ways throughout this period but perhaps their most explicit and comprehensive formulation is found in the Scientific Policy Resolution of 1958, which was drafted and introduced by Nehru in the Indian Parliament. This resolution, which is less a statement of policy than of broad policy objectives, asserted that:

1. to foster, promote, and sustain by all appropriate means, the cultivation of science, and scientific research in all its aspects -- pure, applied and educational;

2. to ensure an adquate supply within the country of research scientists of the higher quality, and to recognize their work as an important component of the strength of the nation;

3. to encourage, and initiate, with all possible speed, programmes for the training of scientific and technical personnel, on a scale adequate to fulfill the country's needs in science and education, agriculture and industry, and defense;

4. to ensure that the creative talent of men and women is encouraged and finds full scope in scientific activity;

5. to encourage individual initiative for the acquisition and dissemination of knowledge, and for the discovery of knowledge, in an atmosphere of academic freedom;

6. and, in general, to secure for the people of the country all the benefits that can accrue from the acquisition and application of scientific knowledge.[10]

This approach to science and science-based technology as instruments in building national power and tackling pervasive social and economic problems has continued to be the dominant thrust of government policy toward science and technology in the two decades since Nehru's death. Perhaps reflecting her father's influence, his daughter, Indira Gandhi, on succeeding him as Prime Minister, maintained continuity in the essential thrust of his efforts to build India's capacity in science and technology as a major source of national political power and as means of trying to meet widespread economic and social needs. Baldev Raj Nayar attributes to her three major accomplishments during her period of power from 1966 to 1984:[11]

1. the expansion and deepening of the science and

technology base;

2. the rationalization of the science and technology system and according a more prominent role to scientists and technologists in decision-making;

3. the significant endeavor to advance research and development outside the government where it was almost exlusively concentrated earlier.

This explicitly instrumental character of Indian science and technology policy has been carefully described in Nayar's monumental work.[12] It is quite clear that Nehru prevailed not only due to his own clearly articulated thinking on the subject but also due to the intervention of certain historical events. The assassination of Gandhi in 1948, the rapid growth of sectarianism within Gandhian ranks that followed, the war with Pakistan and the inability of India to acquire arms from the United States and the Soviet Union in 1947 to fight in Kashmir, subsequent border wars with Pakistan and China, the re-arming of Pakistan during the Eisenhower-Dulles era, the American tilt toward Pakistan during the Nixon-Kissinger era, and the internal challenges to Indian political unity steadily muted any challenge to Nehru's policies from those who advocated a predominantly rural economy and an intermediate but appropriate technology for India.

To implement the policies determined under Nehru and subsequently under Indira Gandhi's prime ministership, institutions such as the Indian Planning Commission and the Atomic Energy Commission, run by eminently qualified Nehru followers, such as P. C. Mahalanobis and Homi Bhabha, were established, and soon became more powerful than most cabinet departments. The rhetoric of "socialist pattern of society" notwithstanding, large industrial houses were encouraged to build and rebuild capital intensive industries with foreign collaboration.

For a brief period, during the prime ministership of Lal Bahadur Shastri (1964) -- a man who described himself as small, with small dreams, and a believer in small projects wich required minimal capital -- there was some hope of replacing the national power bias in Indian science and technology policy with an economic orientation that would meet the basic needs of India's teeming villagers. His untimely death in 1966 put an abrupt end to this embryonic reorientation.

The order established by Nehru was reaffirmed, with new vigor, by India's third prime minister, Indira Gandhi, Nehru's daughter. Objections to this policy were contained by a deliberate policy of providing minimal, but visible, subventions to several show case Gandhian rural projects, and by paying public homage to the Gandhian saints such as land reformer Vinoba Bhave. A public movement for appropriate technology ("Sarvodaya") led by Jayaprakash Narayan, a Marxist-turned-Gandhian-socialist, failed to make any dent in India's well entrenched heavy industries and hi-tech science and technology orientation. Nowhere was this bias in favor of modern and ultra-modern technology more visible than in new institutions for manpower training. Five institutues of technology, patterned after the Massachusetts Institute of Technology, and several regional engineering colleges, agricultural universities, and medical institutes were promoted, with an urban-industrial bias in the curriculum, while science education in villages and towns, elementary and secondary schools lagged badly.

Institutionalizing Indian Capabilities

These engineering and other professional institutions were part of a triad which dedicated itself to enhancing India's research and design capabilities. A network of research institutes, both national and regional in scope, both disciplinary (such as National Physical Laboratory) and interdisciplinary (such as Central Food Research Institute), formed the other part of the triad. Also, massive investments were made in university education and a separate University Grants Commission was empowered to oversee post-secondary education and academic research.

During the approximately four decades of its independence, India enlarged some 40 times the government expenditure on research and development. In the decades of the 1960s and 1970s, the "research ratio" (percentage of gross national product spent on research and development) has almost tripled so that it is now in the neighborhood of 0.6 percent (still well below the leading scientific nations of the world with research ratios of over one percent of GNP but a substantial advance over the meager levels of 0.15 percent and 0.2 percent of earlier years).[13] (See Tables 8.1, 8.2, and 8.3)

The record is, if anything, even more impressive in terms of human resources. R & D personnel have increased many times over during this period. The total stock of scientific and technical personnel has grown well over ten times since 1950 and by 1980 numbered well over two million. (See Tables 8.4 and 8.5) The graduation of scientific and technical personnel from the universities has likewise grown dramatically.[14]

TABLE 8.1
National Expenditure on Research, Development, and Related Scientific and Technological Activities, 1948-49 to 1983-84 (in Rupees crores at current prices)*

Year	1948-49	1950-51	1955-56	1958-59	1965-66	1970-71	1974-75
(A) Expenditure on R&D							
Central Sector	1.10	4.68	12.14	21.78	62.45	112.47	231.14
State Sector	n.a.	n.a.	n.a.	1.00	3.51	12.58	24.00
Private Sector	n.a.	n.a.	n.a.	0.15	2.43	14.59	36.46
Total (A)	1.10	4.68	12.14	22.93	68.39	139.64	291.60
(B) Expenditure on Related S&T Activities							
Central Sector	n.a.	n.a.	n.a.	5.88	16.67	33.73	27.78
State Sector	n.a.	n.a.	n.a.	n.a.	n.a.	n.a.	4.65
Total (B)	n.a.	n.a.	n.a.	5.88	16.67	33.73	32.43
Grand Total (A+B)	1.10	4.68	12.14	28.81	85.06	173.37	324.03

(Continued)

TABLE 8.1 (Continuation)

Year	1976-77	1977-78	1978-79	1979-80	1980-81	1981-82	1982-83	1983-84 Estimated
(A) Expenditure on R&D								
Central Sector	300.54	343.92	412.49	500.36	545.31	721.94	908.67	1044.97
State Sector	25.20	28.50	40.24	46.04	51.95	71.79	88.62	99.25
Private Sector	48.42	58.20	75.87	92.14	103.93	147.00	161.38	193.65
Total (A)	374.16	430.62	528.60	638.54	701.19	940.73	1158.67	1337.87
(B) Expenditure on Related S&T Activities								
Central Sector	9.95	11.61	19.32	23.07	27.97	43.21	54.89	63.12
State Sector	6.85	8.00	10.77	12.72	14.01	19.51	24.00	26.88
Total (B)	16.80	19.61	30.09	35.79	41.98	62.72	78.89	90.00
Grand Total (A+B)	390.96	450.23	558.69	674.33	743.17	1003.45	1237.56	1427.87

Notes: 1. Data for 1948-49 represents only expenditure of Council for Scientific and Industrial Research, Indian Council of Agricultural Research, Indian Council for Medical Research, and Department of Atomic Energy.

2. A number of organizations are engaged in scientific and technological activities, such as weather forecasting, geophysical surveys, teaching consultancy, etc. In addition, they also undertake research for which in a number of cases no separate account is maintained. Wherever such details have not been provided, their expenditure on research has been estimated.

3. The number of units in the private sector varies from year to year. Data for 1965-66 related to 60 companies and that for 1970-71 to 109 companies. From 1975-76 data relates to the companies recognized by the Government of India Department of Science and Technology under the Import Trade Control Policy, for 1975-76 the data relates to 300 companies, for 1978-79 and 1979-80 to 470 companies and for 1980-81 to 1982-83 to 600 companies.

4. Data for 1983-84 has been estimated by applying the following rates of growth: Central Sector 15%, State Sector 12%, Private Sector 20%.

* A crore is RS. 10,000,000 (approximately $830,000 at current exchange rates which, however, have fluctuated substantially during the period covered by the table.

Source: Government of India, Department of Science and Technology, *Research and Development Statistics, 1982-83*, New Delhi: DST, 1984.

TABLE 8.2
Expenditure on Science and Technology in Relation to Gross National Product, 1958-59 to 1982-83
(Monetary Amounts in Rs. Crores at Current Prices)*

Particulars	1958-59	1965-66	1970-71	1973-74	1974-75	1976-77
GNP at current prices (Rs. crores)	12600	21866	36452	53447	62972	71575
Expenditure on R&D (Rs. crores)	22.93	68.39	139.64	216.01	291.60	374.16
Expenditure on R&D as percentage of GNP	0.18	0.31	0.38	0.40	0.46	0.52
Expenditure on R&D and related S&T activities (Rs. crores)	28.81	85.06	173.37	253.53	324.03	390.96
Expenditure on R&D and related S&T activities as percentage of GNP	0.23	0.39	0.47	0.47	0.52	0.55

(Continued)

TABLE 8.2 (Continuation)

Particulars	1977-78	1978-79	1979-80	1980-81	1981-82**	1982-83**
GNP at current prices (Rs. crores)	80946	86754	94052	112156	131740	145141
Expenditure on R&D (Rs. crores)	430.62	528.60	638.54	701.19	940.73	1158.67
Expenditure on R&D as percentage of GNP	0.53	0.61	0.68	0.62	0.71	0.80
Expenditure on R&D and related S&T activities (Rs. crores)	450.23	558.69	674.33	743.17	1003.45	1237.56
Expenditure on R&D and related S&T activities as percentage of GNP	0.56	0.64	0.71	0.66	0.76	0.85

* A crore is RS. 10,000,000 (approximately $830,000 at current exchange rates which, however, have fluctuated substantially during the period covered by the table).
** Estimated

Source: A. Rahman, *Science and Technology in India*, New Delhi: National Institute of Science, Technology and Development Studies (NISTADS), 1984; Government of India, Department of Science and Technology, *Research and Development Statistics, 1982-83*, New Delhi: DST, 1984.

TABLE 8.3

Research and Development Expenditure by Major Scientific Agencies of the Government of India (in Rs. Lakhs at Current Prices)*

Name	R&D Expenditure (Rs. Lakhs)								
	1958-59	1969-70	1973-74	1976-77	1978-79	1979-80	1980-81	1981-82	1982-83
Department of Atomic Energy	775.88	2072.06	2445.63	5831.74	6082.32	6781.46	7623.24	8826.06	10563.13
Council of Scientific and Industrial Research	509.94	1868.07	2500.77	4125.71	5592.39	5918.99	7281.79	7877.08	10081.50
Defense Research and Development Organization	150.00	1454.39	3428.97	5065.00	6678.61	9662.91	7970.00	10483.36	12199.96
Indian Council of Agriculture Research	372.29	1377.08	2408.42	3739.24	5603.56	7739.81	6599.27	11149.50	13121.50
Indian Council of Medical Research	50.50	156.46	170.00	423.00	533.54	598.50	842.08	1195.13	1423.15
Department of Science and Technology	22.13	73.88	213.67	1121.20	2389.83	2603.56	3537.30	4608.01	6051.06
Department of Space	--	--	1909.22	3892.45	4553.74	4421.13	5270.85	7503.44	8605.17
Department of Electronics	--	--	97.53	422.01	511.47	750.25	431.35	751.11	847.55
Total	1880.74	7001.94	13174.21	24620.35	31945.46	38476.61	39555.88	52393.69	62892.32

* A Lakh is Rs. 100,000 (approximately $8,300 at current exchange rates which, however, have fluctuated substantially during the period covered by the table).

Source: Government of India, Department of Science and Technology, Research and Development Statistics, 1982-83, New Delhi: DST, 1984.

TABLE 8.4
Stock of Scientific and Technical Personnel (1950-1985)

Category of Personnel	Stock at the end of the year (000)						
	1950	1955	1960	1965	1970	1980 Estimated	1985 Estimated
Engineering and Technology							
Degree	21.6	37.5	62.2	106.7	185.4	221.4	266.3
Diploma	31.5	46.8	75.0	138.9	244.4	329.4	429.9
Science							
Post Graduates	16.0	28.0	47.7	85.7	139.2	217.5	273.0
Graduates	60.0	102.9	165.6	261.5	420.0	750.3	956.5
Agriculture							
Post Graduates	1.0	2.0	3.7	7.7	13.5	96.5	114.1
Graduates	6.9	11.5	20.2	39.4	47.2	--	--
Medicine							
Degree	18.0	29.0	41.6	60.6	97.8	167.6	198.7
Licentiate	33.0	35.0	34.0	31.0	27.0	N.A.	N.A.
Total	188.0	292.7	450.0	731.5	1174.5	1782.7	2238.5

Source: A. Rahman, _Science and Technology in India_, New Delhi: National Institute of Science, Technology and Development Studies (NISTADS), 1984.

TABLE 8.5
Personnel Employed in Research and Development Institutions in India, 1982

Name of Establishment	Personnel engaged primarily in R&D activities	Personnel engaged in auxiliary/ scientific/ technical activities	Personnel employed in administrative and other non-technical supporting activities	Total
Department of Atomic Energy	5199	7076	5092	17367
Council of Scientific & Industrial Research	6973	10220	6291	23484
Defense Research & Development Organization	6904	7131	8213	22248
Indian Council of Agricultural Research	6222	5402	7861	19485
Department of Space	3236	4310	3600	11146
Indian Council of Medical Research	741	839	871	2451
Other Central Government Ministries/Agencies	14909	14947	15699	45555
State Governments	13411	5227	10284	28922
Private Sector	14003	5735	6499	26237
Total	71598	60887	64410	196895

Source: Government of India, Department of Science and Technology, *Research and Development Statistics, 1982-83*, New Delhi: DST, 1984.

In his authoritative work, Nayar sketches in the following "broad profile" of the science and technology systems that had emerged by 1980 from the various government policies and political actions pursued since independence and mentioned above:[15]

119 universities, affiliating about 16,500 colleges
5 institutes of technology
150 engineering colleges
100 medical colleges
350 polytechnics
150,000 annual addition to the total stock of qualified scientific and technical personnel
2.5 million as a total stock of qualified scientific and technical personnel, the third largest in the world
130 specialized research laboratories and institutes under the auspices of the Indian Council of Agricultural Research, Council of Scientific and Industrial Research, Indian Council of Medical Research, Department of Space, and Defense, Research and Development Organization
600 in-house R & D laboratories in the public sector and private sector industrial enterprises
150 engineering consultancy organizations of varying sizes, employing 20,000 technologists
0.6 percent as the share of total expenditure on science and technology in GNP

Budgetary allocations for this massive science and technology effort were planned by the all powerful Indian Planning Commission, every five years, until recently. When one reviews the reports of the Commission, one finds that the approach of the planners was both pragmatic and haphazard. Allocations were often determined, and certainly modified, by the requirements of Indian federalism. Colleges, universities, research institutions, and national and regional laboratories were distributed among various states of the Union as political pork barrel.

The personality factor was important. Individual scientists with political connections were quite successful. Take for example physicist Homi Bhabha, the father of India's nuclear effort, and S. Hussain Zaheer, the colorful director of the Central Council on Scientific and Industrial Research, with their connections to the Nehru family. Or take Y. Nayudamma, the creator of the Central Leather

Research Institute and the support he received from Kamaraj Nayar, the powerful boss of the ruling Congress Party. There was nothing sinister in this. So much was to be done in India, so much remained to be done on the sub-continent, that any scientist with the right credentials and appropriate political support could create an institution for manpower development and research investigations and design.

There was a parallel political process for establishing an institution which had or was about to become self-sustaining in administrative structure, instead of being dependent on a particular personality. The Indian Institute of Science and the University of Roorkee are cases in point.

The approach of the planners as well as of the funding agencies has been truly protean. Yet, needless to say, in the absence of clear rationalizations for allocations, there has also been much waste. Much of this waste was made possible by the availability of significant government revenues through a very high rate of internal savings in India and also from massive external assistance, particularly from the World Bank and the United States (PL 480 funds).

No doubt about it: India has produced a substantial force of scientific personnel. Graduates of the foreign sponsored institutes of technology, the All India Institute of Medical Sciences, the Indian Institute of Science, and the G. B. Pant Agricultural University, among others, are as good as any produced by the leading institutions in the West. Graduates of regional engineering colleges, regional medical and agricultural institutes, and other established professional institutions are certainly at par with the graduates of western colleges and universities that do not offer doctoral degrees. All this is visible through Indian educated engineers and physicians now settled in the West.

This quality training of scientists in India cannot mask two painful facts. The first is that India has not as yet been transformed into a technologically oriented community. Second, there is a massive unemployment of technical and scientific personnel, with about 12 percent of recent graduates unable to find productive employment within three years of their graduation. This situation would be worse if the government were not the employer of about 65 percent of all engineering graduates.[16] Appropriate linkages between training institutions and potential employers have not developed. Each annual report of the Council on Scientific and Industrial Research laments this fact. Consequently those who can leave the Indian job market do so by emigrating to the West, the oil producing countries, and the English speaking countries of Africa with an inadequate supply of engineering and technical personnel. In spite of the abundance of engineers and other professionals in India, school graduates still aspire for admission to professional curricula largely because liberal arts education offers even bleaker prospects.

The Results Of Capacity Building

Our criticisms notwithstanding, much has accomplished in India. India has used science and technology effectively as a means of enhancing its self-sufficiency in arms and weapons. Even if its entry into the nuclear field was greeted with something less than enthusiasm by the existing members, it has joined the elect membership of the nuclear club with its nuclear explosion in May, 1974.

Impressive progress has also been made by India in its goal of becoming a major industrial nation. It ranks between eighth and tenth among the nations of the world by most indicators of industrial capacity. It has greatly expanded its steel, chemical, and mechanical engineering industries. It produces a wide variety of basic manufacturing goods and has begun to export technology as well as technology-intensive goods and services such as engineering design consultancy and computer software.[17]

Nor is that the end of the list. The nuclear generation of electric power, a capital goods industry in power generation and many other fields, and an electronics industry have also become part of the Indian industrial landscape. Equally significant changes have occurred in other sectors of economic and social life, including agriculture (where the Green Revolution has connected many of India's larger farmers with access to resources to a chemical and energy-intensive technology) and in public health.[18] India is now an exporter of agricultural products.

We could go on but the basic point should by now be clear. Capacity building in science and technology, to which independent India has made a substantial commitment over the past four decades, has played an important role in the enhancement of national political power, including the development of a modern industrial base. These developments have been amply described elsewhere and need not be elaborated here.[19] (Tables 8.4, 8.5) Morehouse and Chopra, 1983; Morehouse, Gupta and Deolalikar, 1980; Rahman, 1984;

Rahma, 1985.

We are not suggesting that India has become competitive with the West or self-sufficient in industrial design, technological development, and professional management. For many large scale plants, such as the tragic Union Carbide facility in Bhopal, designs are still commissioned abroad. However, Indian ability to assess competing advanced technologies stands at a high level. Its ability to make incremental improvements and modifications to designs obtained from overseas is second to none. Inadequacy of utilities (power, water, industrial waste disposal, and telecommunications), not money or manpower, are holding India back. Even without the benefit of the revolution wrought by microelectronics, Indian capabilities in day-to-day technical management are good. Indian scientific and technical development has become a model for developing non-communist nations, and the Indian technical genius is now at work in many foreign countries, both developed and developing.

The Impact On Society

Yet the realization of a socialistic and more egalitarian socialist pattern of society has remained as elusive as ever. The position of the rural poor has deteriorated.[20] Small and marginal farmers have been removed from their land because of their inability to invest in modern inputs such as chemicals.[21] It is unlikely that the problems of a vast number of villagers could have been overcome in a substantial measure in such a short period of time as four decades, no matter what strategies were pursued. Yet it is also true that the relentless pursuit of science and technology for national power has not mitigated the predicament of the villagers.

It is important to avoid the luddite trap. There is abundant evidence that due to electricity, better transportation, increased agricultural productivity, better medical and public health, and public education, there is discernible improvement in the Indian quality of life. Life expectancy has improved from about 32 years in 1950 to about 52 years in 1981.[22] India is now an exporter of cereals and food-grains. Yet, as Kurien's study of Tamil Nadu points out, the gains of science and technology have been most unevenly distributed, and social inequalities have increased.[23]

In India, as elsewhere, the debate continues on the

relevance of "appropriate" technology in the modern world, a technology that is primarily based on a country's human factors and other endowments. Maximization of employment has never been the top priority in Indian economic planning, and consequently research and design of appropriate technology has not received either political direction or adequate resource funding. Proponents of appropriate technology have not become a political force in India, and the break-throughs that many individual designers of appropriate technology have achieved have been generally ignored.

In one study, Bhalla showed that the capital output ratio for traditional methods of spinning using Ambrachakra craft models were lower than factory models.[24] A. K. Sen pointed out that in cotton-weaving the capital output ratio was the lowest in the fly-shuttle handloom, a labor intensive method, than in the automatic power loom. G. K. Boon had provided evidence on the existence of a range of efficient techniques in a number of industries below a certain critical scale of output.[25] Makhajani has pointed out the greater efficiency of alternative, labor-intensive, methods of producing energy for villagers, and A.K.N. Reddy has consistently written about the economic viability of bio-gas and its positive impact on village life and employment.[26] Frances Stewart and Paul Streeten (1971) have come to the conclusion that there is no conflict between output and employment objectives in most developing countries.[27] Even though all students of Indian political economy agree with Gunnar Myrdal that "the hope so commonly expressed that a large proportion of those who will join the labor force in the decades to come will become productively employed outside agriculture is illusory," there is neither a serious drive nor any political commitment to appropriate technology in India.[28]

Proponents of appropriate technology have united behind a common blueprint. It calls for the scaling down of large scale manufacture of (1) sugar, (2) cement, (3) paper, (4) cotton and jute spinning and weaving, (5) fertilizers, (6) soap making, and (7) other day-to-day consumption goods. They have called for upgrading of village technologies in (1) hand-loom weaving, (2) extraction of vegetable oils, (3) tanning and shoe-making, (4) cereal processing, and (5) many other traditional village crafts.

This is not the end of the list. Bio-gas utilization, greater use of tube wells, village based cooperative production of poultry and dairy products, and refinement of traditional methods of waste disposal have all been recommended. The ideas are all there. So is the village constituency, but not the political movement necessary to make it happen on a substantial scale. To most Indians close to the levers of political power, appropriate technology remains a romantic vision, a grand mirage.

Summing Up: Four Decades Of Capacity Building

Since India began its "tryst with destiny," to use Nehru's phrase, in August 1947, enormous strides have been taken in strengthening its science and technology capabilities. During the four decades since independence, Indian has -- building upon a substantial pre-independence base, including an important indigenous tradition in science and technology -- emerged as a significant figure on the world science and technology map. It is clearly among the leading Third World countries -- in a class probably only with China and Brazil -- in what it has achieved during this period. In short, it has effectively institutionalized a complex and extensive infrastructure for the doing of modern science and science-based technology.

India has thus made substantial progress toward achievement of one of the two basic objectives in its effort to build its capacity in science and technology -- to eliminate -- or at least diminish -- dependent relationships with the industrialized North by strengthening its modern industrial base and its military capabilities.

But it has made precious little progress in pursuit of the other basic objective of meeting the rising expectations of its people in meeting their most fundamental requirements for food, shelter, health, education, and employment, even though there has been significant impact on all of these aspects of life. What the Indian experience makes abundantly clear is that deep-rooted social problems such as poverty and exploitation cannot be solved through technical fixes such as institutionalizing indigenous capability in science and technology.

NOTES

Parts of this chapter have been adapted from Ward Morehouse, "Myth and Reality: Animadversions on Science, Technology, and Society in India," Knowledge: Creation, Diffusion, Utilization, June 1985. Vol. 6, No. 4, and "The

Mythology of Modernity: Science, Technology, and Social Change in India," in B. N. Varma, ed., India's Modernization: Tradition and Change, forthcoming.

1. Claude Alvares, Homo Faber: Technology and Culture in India, China and the West, 1500 to the present day (New Delhi: Allied Publishers, 1979). H. T. Bernstein, Steamboats on the Ganges: An Exploration in the History of India's Modernization Through Science and Technology (Calcutta: Orient Longman, 1960).

2. George Basalla, "Spread of Western Science", Science, May 5, 1967, pp. 617-688.

3. S. Abid Husain, The Way of Gandhi and Nehru (London: Oxford University Press, 1959), p. 45-46.

4. J. C. Kumarappa, Why the Village Movement? Wardha: Village Industries Association, 1948) and Bhartan Kumarappa, Capitalism, Socialism or Villagism, foreword by M. K. Gandhi, (Varanisi: Sarva Seva Sangh), 2nd. ed. 1965.

5. Dorothy Norman, ed., Nehru, The First Sixty Years, 2 Vols. (New York: Columbia University Press, 1965), Vol. II, p. 179.

6. Jawaharlal Nehru, The Discovery of India (New York: John Day, 1946), p. 411.

7. Ibid. i.

8. Ibid., p. 26.

9. Ward Morehouse, "Nehru and Science: The Vision of New India," Indian Journal of Public Administration 15, 1969, pp. 489-508.

10. A. Rahman, et al. Science and Technology in India (New Delhi: Indian Council for Cultural Relations, 1973) contains the complete text of the 1958 Resolution.

11. Baldev Raj Nayar, India's Quest for Technological Independence, 2 Vols. (New Delhi: Lancers, 1983).

12. Ibid. Vol. I, p. 411.

13. Ministry of Science and Technology, Government of India, Science and Technology in India: Retrospect and Prospect (New Delhi: DST, 1985), and Research and Development in Industry: 1982-83 (New Delhi: DST, 1984), and Research and Development Statistics, 1982-83 (New Delhi: DST, 1984).

14. Ward Morehouse, Science in India: Institution Building and the Organizational System for Research and Development (Bombay and Hyderbad: Administrative Staff College of India and Popular Prakashan, 1971) and Baldev Raj Nayar, op. cit.

15. Nayar, op. cit. pp. 537-538. Vol. I.

16. Ministry of Information and Broadcasting, India: A Reference Annual (New Delhi: Publication Division, 1981).

17. Joseph N. Grieco, *Between Dependency and Autonomy: India's Experience with the International Computer Industry* (Berkeley: University of California Press, 1984).

18. Indian Agricultural Research Institute, *Green Revolution in India* (New Delhi: The Indian Investment Centre, 1970). G. S. Bhalla and G. K. Chadra, *Green Revolution and the Small Peasant: A Study of Income Distribution in Punjab Agriculture* (New Delhi: Centre for the Study of Regional Development, Jawaharlal Nehru University, 1982). C. T. Kurien, *Dynamics of Rural Transformation: A Study of Tamil Nadu 1950-1975* (New Delhi: Orient Longman, 1981).

19. A representative selection out of a vast literature includes Nayar, op. cit. Vol. 2; Aqueil Ahmad, "Scientific Community in India: An Illusive Trademark," *Bulletin of Science*, New Delhi, April-May 1985, and "Politics of Science Policy Making in India," *Science and Public Policy*, London, October 1985; A. Rahman, *Science and Technology in India* (New Delhi: National Institute of Science, Technology, and Development Studies, 1984), and *Science and Technology in Indian Culture: An Historical Perspective* (New Delhi: National Institute of Science, Technology, and Development Studies, 1984). Ward Morehouse and Ravi Chopra, *Chicken and Egg: Electronics and Social Change in India* (Lund: Research Policy Institute, University of Lund, 1983). Ward Morehouse, Brijen K. Gupta, Anil Deolalikar, *Assessment of U.S.-Indian Science and Technology Relations: An Analytical Study of Past Performance and Future Prospects* (Springfield, VA: National Technical Information Service, 1980).

20. International Labour Office, *Poverty and Landlessness in Rural Asia* (Geneva: ILO, 1977). M. D. Morris, *Measuring the Condition of the World's Poor* (New York: Pergamon, 1979).

21. Claude Alvares, "Development Against People," *Development Forum*, New York, 1978.

22. Ministry of Health and Family Welfare, Government of India, *Family Welfare Programmes in India Yearbook 1981-1982*, New Delhi.

23. C. T. Kurien, *Dynamics of Rural Transformation: A Study of Tamil Nadu 1950-1975* (New Delhi: Orient Longman, 1981), pp. 145-148.

24. A. S. Bhalla, "Investment Allocation on Technological Choice: A Case of Cotton Spinning Techniques," *Economic Journal*, London, 1964.

25. G. K. Boon, *Economic Choice of Human and Physical Factors in Production* (North Holland: Elsivier, 1964).

26. A. Makhajani, *Energy and Agriculture in the Third*

World (New York: Ford Foundation, 1975). C. R. Prasad, et. al., "Biogas Plants," Economic and Political Weekly, Bombay, special number, 1974.

27. Frances Stewart and Paul Streeten, "Conflicts between Output and Employment Objectives," Oxford Economic Papers, 1971.

28. Gunnar Myrdal, Asian Drama, 3 Vols. (New York: Twentieth Century Fund, 1968), Vol. 3, p. 412.

Index